腔光力学系统中的
量子光学效应及应用

江 成 著

清华大学出版社
北京

版权所有，侵权必究。举报：010-62782989，beiqinquan@tup.tsinghua.edu.cn。

图书在版编目(CIP)数据

腔光力学系统中的量子光学效应及应用/江成著.—北京：清华大学出版社，2019 (2024.11重印)
　　ISBN 978-7-302-53622-2

　　Ⅰ.①腔… Ⅱ.①江… Ⅲ.①量子光学-光学效应-研究 Ⅳ.①O431.2

中国版本图书馆 CIP 数据核字(2019)第 173905 号

责任编辑：鲁永芳
封面设计：常雪影
责任校对：赵丽敏
责任印制：宋　林

出版发行：清华大学出版社
　　　　　网　　址：https://www.tup.com.cn，https://www.wqxuetang.com
　　　　　地　　址：北京清华大学学研大厦A座　邮　　编：100084
　　　　　社 总 机：010-83470000　　　　　　　　邮　　购：010-62786544
　　　　　投稿与读者服务：010-62776969，c-service@tup.tsinghua.edu.cn
　　　　　质量反馈：010-62772015，zhiliang@tup.tsinghua.edu.cn
印 装 者：三河市人民印务有限公司
经　　销：全国新华书店
开　　本：170mm×240mm　　印　张：12.25　　字　数：187千字
版　　次：2019年12月第1版　　　　　　　印　次：2024年11月第3次印刷
定　　价：79.00元

产品编号：084839-01

前　言

　　腔光力学是最近十多年来进展十分迅速的一个热门研究领域,主要研究光学微腔或者微波腔与微纳米机械振子通过辐射压力耦合形成的相互作用。早在 20 世纪 60 年代,布拉金斯基(Braginsky)等就研究了辐射压力对机械振子的动态影响。最近的实验和理论进展表明,该领域不但为研究宏观机械物体的量子力学效应提供了一个很好的平台,而且在经典和量子信息处理以及精密测量方面(比如力、位移、质量测量等)有着重要的应用。腔光力学系统中腔场与机械振子之间的相互作用主要产生两方面的影响。一方面,利用腔场可以对机械运动进行量子调控和冷却。最近,微波和光学领域的腔光力学系统中的纳米机械振子实验上已被成功地冷却到量子基态,从而为腔光力学系统中观察到各种量子力学效应铺平了道路。另一方面,与机械振子的相互作用会改变腔场的光学响应,产生一系列新颖的量子光学现象。当腔场受到红失谐的泵浦场驱动时,系统中会出现光力诱导透明现象,从而使腔光力学系统在快慢光以及共振增强的非线性效应等方面有重要的应用前景。当腔场受到蓝失谐的泵浦场驱动时,系统中可出现光力诱导吸收和放大现象,可以有效地放大弱的光或者微波信号,因此在通信网络中有着潜在的应用。此外,当腔场受到共振的泵浦场驱动时,光力系统可用于位移、质量等精密测量中。

　　本书内容主要依托作者近年来在腔光力学系统中的研究成果,共分为 6 章。第 1 章绪论,主要介绍腔光力学系统的研究背景、基本理论、几种典型的腔光力学系统、光力诱导透明、光力诱导吸收和放大等。第 2 章分别介绍双腔光力系统、受二能级调制的混杂光力系统以及宇称-时间-对称的光力系统中的慢光效应,主要讨论腔场受到红失谐的泵浦场驱动时系统中出现的光力诱导透明和法诺(Fano)共振等,并进一步研究系统中的慢光效应。第 3 章分别介绍双腔光力系统、二能级系统耦合腔场形成的混杂光力系统,以及二能级系统耦合机械振子形成的混杂光力系统中的光学双稳态，研究稳态时腔内光子数和机械振子的声子数等随泵浦场频率和功率等因素变化时出现的双稳现象。第4章介绍两种不同的光力系统在受到蓝失谐的泵浦场驱动时探测场透射谱中出现的光力诱导吸收和放大现象，研究泵浦场的频率和功率以及机械驱动的振幅和相位等因素对系统光学响应的影响。第5章介绍由一个微波腔和一个光腔耦合于一个共同的机械振子形成的混杂光-电力系统在受到共振的泵浦场驱动时在超灵敏质量传感器中的应用，研究如何根据探测场透射谱及四波混频谱中共振频率峰值的移动进行纳米颗粒、病毒等微小物质的质量测量。第6章介绍由三个腔场耦合于一个共同的机械振子形成的三腔光力系统在单向放大器中的应用，该系统中的相互作用构成了一个闭合回路，因此回路的整体位相差可用于控制系统中的非互易性传输特性。6.1节和6.2节分别介绍通过引入光学增益和施加蓝失谐的泵浦场这两种方式实现相位保持的单向放大器，6.3节介绍通过同时施加红、蓝失谐的泵浦场驱动实现相位敏感的单向放大器。

全书彩图请扫右侧二维码。

由于作者水平有限，书中难免有差错和疏漏之处，恳请读者批评指正。

在本书成书过程中，淮阴师范学院崔元顺教授、陈贵宾教授以及北京计算科学研究中心李勇研究员给予了大力支持与帮助，作者表示衷心感谢！本书受到了江苏省"青蓝工程"的部分资助。

江　成

淮阴师范学院

2019年4月

目　　录

第 1 章　绪论 ··· 1

　1.1　背景介绍 ··· 1

　1.2　腔光力学系统的基本理论 ·· 3

　　　1.2.1　光学微腔 ··· 3

　　　1.2.2　机械振子 ··· 5

　　　1.2.3　光力耦合 ··· 6

　1.3　几种典型的腔光力学系统 ·· 8

　　　1.3.1　悬挂的镜子 ·· 9

　　　1.3.2　光学微振子 ·· 10

　　　1.3.3　波导和光子晶体腔 ··· 10

　　　1.3.4　悬浮的纳米物体 ·· 11

　　　1.3.5　微波振子 ··· 12

　　　1.3.6　超冷原子 ··· 12

　1.4　腔光力学系统的光学响应 ·· 13

　　　1.4.1　光力诱导透明 ··· 13

　　　1.4.2　光力诱导吸收和放大 ······································ 16

　参考文献 ··· 16

第 2 章　腔光力学系统中的慢光效应 ································ 22

　2.1　双腔光力系统中的光力诱导透明和慢光效应 ················· 22

　　　2.1.1　引言 ··· 22

　　　2.1.2　模型和理论 ·· 23

　　　2.1.3　结果和讨论 ·· 26

2.1.4　小结 ·· 29
　　　参考文献 ··· 29
2.2　混杂光力系统中受二能级系统调制的法诺共振
　　和慢光效应 ·· 32
　　　2.2.1　引言 ·· 32
　　　2.2.2　模型和理论 ··· 33
　　　2.2.3　可控的法诺共振 ·· 36
　　　2.2.4　探测场透射谱中的慢光效应 ··· 40
　　　2.2.5　小结 ·· 42
　　　参考文献 ··· 43
2.3　机械驱动下宇称-时间-对称的光力系统中的快慢光效应 ········· 47
　　　2.3.1　引言 ·· 47
　　　2.3.2　模型和理论 ··· 48
　　　2.3.3　机械驱动调制的探测场透射谱 ····································· 53
　　　2.3.4　透射探测场中可控的慢光和快光效应 ·························· 57
　　　2.3.5　小结 ·· 60
　　　参考文献 ··· 60

第3章　腔光力学系统中的光学双稳态 ···································· 64

3.1　双腔光力学系统中可控的光学双稳态 ···································· 64
　　　3.1.1　引言 ·· 64
　　　3.1.2　模型和理论 ··· 65
　　　3.1.3　结果和讨论 ··· 66
　　　3.1.4　小结 ·· 70
　　　参考文献 ··· 71
3.2　二能级原子与腔场耦合的混杂光力系统中的光学双稳态和
　　动力学效应 ·· 73
　　　3.2.1　引言 ·· 73
　　　3.2.2　模型和理论 ··· 75

3.2.3 光子数和布居数反转的双稳态行为 …………………… 78
3.2.4 初始条件和腔泵浦强度对系统动力学效应的影响 …… 80
3.2.5 小结 ………………………………………………… 83
参考文献 …………………………………………………… 83

3.3 二能级原子与机械振子耦合的混杂光力系统中的
光学双稳态和四波混频 ……………………………………… 86
3.3.1 引言 ………………………………………………… 86
3.3.2 模型和理论 ………………………………………… 87
3.3.3 光子数和声子数的双稳行为 ……………………… 90
3.3.4 共振增强的四波混频过程 ………………………… 92
3.3.5 小结 ………………………………………………… 96
参考文献 …………………………………………………… 96

第4章 腔光力学系统中的光力诱导吸收和放大 …………………… 100

4.1 机械驱动下多模光力系统中相位依赖的光力诱导吸收 ……… 100
4.1.1 引言 ………………………………………………… 100
4.1.2 模型和理论 ………………………………………… 101
4.1.3 结果和讨论 ………………………………………… 105
4.1.4 小结 ………………………………………………… 111
参考文献 …………………………………………………… 112

4.2 混杂光力系统中可控的光学响应 ……………………………… 115
4.2.1 引言 ………………………………………………… 115
4.2.2 模型和理论 ………………………………………… 116
4.2.3 数值结果和讨论 …………………………………… 119
4.2.4 小结 ………………………………………………… 122
参考文献 …………………………………………………… 122

第5章 基于混杂光-电力系统的超灵敏纳米机械质量传感器 ……… 126

5.1 引言 ………………………………………………………… 126

5.2 模型和理论 …………………………………………………… 127
5.3 结果和讨论 …………………………………………………… 131
5.4 小结 …………………………………………………………… 135
参考文献 …………………………………………………………… 135

第6章 基于三腔光力系统的单向放大器 …………………………… 139

6.1 基于包含增益的三腔光力系统的单向放大器 ………………… 139
    6.1.1 引言 ……………………………………………………… 139
    6.1.2 模型 ……………………………………………………… 140
    6.1.3 单向放大器 ……………………………………………… 144
    6.1.4 慢光效应 ………………………………………………… 150
    6.1.5 小结 ……………………………………………………… 151
    参考文献 ……………………………………………………… 152

6.2 基于三腔光力系统的量子极限的单向放大器 ………………… 156
    6.2.1 引言 ……………………………………………………… 156
    6.2.2 模型和理论 ……………………………………………… 157
    6.2.3 量子极限的单向放大器 ………………………………… 161
    6.2.4 小结 ……………………………………………………… 167
    参考文献 ……………………………………………………… 167

6.3 微波和光学光子之间相位敏感的单向放大器 ………………… 171
    6.3.1 引言 ……………………………………………………… 171
    6.3.2 模型和理论 ……………………………………………… 172
    6.3.3 相位敏感的单向放大器 ………………………………… 177
    6.3.4 小结 ……………………………………………………… 182
    参考文献 ……………………………………………………… 182

第 1 章 绪 论

1.1 背景介绍

光具有动量,当遇到物体被反射时可以产生辐射压力。早在 17 世纪开普勒就注意到彗星运动时其尘埃尾巴的方向总是背离太阳,这种现象与辐射压力有关。光能辐射计实验中首次证实了辐射压力的存在[1],并且实验中需要做一些仔细的分析来区分辐射压力效应和热效应。爱因斯坦于 1909 年得到了作用在可移动镜子上的辐射压力涨落的统计数据[2],其中包括辐射压力的摩擦效应。爱因斯坦通过这个分析揭示了黑体辐射的波粒二象性本质。在早期的实验中,弗里希(Frisch)[3]和贝丝(Beth)[4]分别观察到了光子的线性动量和角动量转移到原子和宏观物体上的现象。

20 世纪 70 年代,阿斯金(Ashkin)证实聚焦激光束可以用来束缚和控制电介质颗粒以及反馈冷却[5]。辐射压力非守恒的特性可能被用来冷却原子运动首先由汉斯(Hänsch)和肖洛(Schawlow)[6]以及维因兰德(Wineland)和德默尔特(Dehmelt)[7]提出。激光冷却随之在 20 世纪 80 年代的实验中被实现并成为一种非常重要的技术,它可以使离子冷却到其运动基态并且是超冷原子实验中潜在的资源。激光冷却使得很多应用成为可能,包括光学原子钟、引力场的精密测量和受限原子云中量子多体物理的系统研究等[8]。

布拉金斯基(Braginsky)在干涉仪中研究了辐射压的作用并发现可以用来冷却较大的物体,他考虑了辐射压对腔中一个简谐振动的悬挂镜子的动态影响,如图 1.1.1 所示。他的分析揭示了腔的有限寿命导致辐射压力的延迟本质可以引起机械运动的阻尼或反阻尼效应,并且这两个效应可以在早期的微波腔实验中被观察到[9]。在之后的实验中,这些现象在微波耦合的千克尺度的机械振子中也被观察到了。在光学领域,第一个腔光力学实验证实了作用在宏观镜子上的辐射压力的双稳性[10]。布拉金斯基还证实,辐射压的量子涨落会导致自由物体位置测量的精密度有一个极限值[11]。卡维斯（Caves）、杰克尔（Jaekel）和雷诺（Renaud）等的详细分析澄清了干涉仪中这种量子噪声的影响[12]。这些工作建立了连续位置探测中的标准量子极限，对激光引力波天文台（LIGO）和处女座干涉仪（VIRGO）等引力波探测器非常重要。

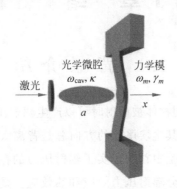

图 1.1.1　典型的腔光力学系统示意图。法布里-珀罗腔的左边镜子固定不动，而右边镜子在辐射压作用下可在平衡位置附近来回振动
（请扫Ⅱ页二维码看彩图）

20世纪90年代，科学家开始对量子腔光力学系统中光的压缩[13]和光强的量子非破坏测量[14]等几个方面进行理论研究，主要是利用由光力相互作用产生的有效克尔非线性。当光力耦合非常强时，产生的量子非线性可以导致光场和机械振子的非经典态和纠缠态。此外，辐射压的反馈冷却被提出[15]。几乎在同一时间，腔辅助的激光冷却作为一种冷却原子和分子运动的方法[16]被提出。

在实验方面，基于辐射压力的光学反馈冷却首先被科哈顿（Cohadon）等在宏观镜子的振动模式实验中实现[17]。这个方法后来发展到温度更低的情况。与此同时，另一个趋势是机械振子的尺寸变得越来越小，比如，毫米尺寸的镜子的热运动在一个低温光学腔中被监测到[18]。但是，制作出低于毫米尺度的高品质法布里-珀罗腔是非常具有挑战性的。尽管如此，在微米尺度的装置中观察到延迟辐射压的光力效应仍然是可能的，其中力是光热起源，可以用热时间常数有效地代替腔的寿命。具体的例子包括光学弹簧效应[19]、反馈阻尼[20]、自持续振荡[21]和腔冷却[22]等。

就量子相干的光力学中的应用而言，最好能够利用非耗散的辐射压力。

光学微腔和先进纳米制造技术的出现使得人们进入这一领域变得可能。2005年人们发现高品质的光学微环型振子可以同时包括力学模式和光学模式,因此可以实现光力效应,尤其是辐射压诱导的自振荡[23]。2006年,三个不同的研究小组分别实现了基于悬挂微镜子和微型环芯的辐射压冷却[24-26]。从此以后,腔光力学领域进展迅速,并且光力耦合在大量新颖的系统中被实现,包括薄膜和纳米棒置于法布里-珀罗腔、回音壁微盘和微球、光子晶体以及耦合的纳米梁等。此外,腔光力学可以利用冷原子云的集体激发实现[27]。光力相互作用在光学波导中也存在,最先用来研究和观察压缩,其中光纤受限的力学模式产生被引导的声波散射[28]。现在有很多系统可以在没有腔时实现光力相互作用,比如光子电路中的波导或者光子晶体光纤[29]。

利用微加工的超导振子也可以实现光力相互作用,包括将一个纳米机械振子嵌入到一个超导传输线微波腔中[30],或者将一个可移动的铝薄膜集成到一个超导振子中[31]。在这些系统中,机械运动电容耦合于微波腔。这个方法使腔光力学可以利用电或者其他的耦合技术来测量和控制纳米机械振子的运动,包括机械振子耦合于单电子晶体管[32]或者量子点接触[33]等。

人们对腔光力学领域产生日益增长的研究热情有以下几个原因:一方面,该领域可以用来实现微小力、位移、质量、加速度等高精度光学探测;另一方面,量子腔光力学为利用光来操控和探测量子区域的机械运动提供了平台,从而产生光和机械运动的非经典态。这些工具为量子信息处理中的应用奠定了基础,其中光力器件可以作为相干的光-物质界面将存储在固态量子比特中的信息转化到飞行的光量子比特中。此外,腔光力学系统可以用来连接不同物理系统的自由度构成混杂的量子器件。

1.2 腔光力学系统的基本理论

1.2.1 光学微腔

首先考虑由两个相距为 L 的高反射率镜子构成的法布里-珀罗腔。腔中存在一系列纵模,角频率为 $\omega_{\text{cav},m} \approx m\pi(c/L)$,其中 m 为整数。两个纵模之间的间隔表示为腔的自由光谱范围:

$$\Delta\omega_{\text{FSR}} = \pi \frac{c}{L} \tag{1.2.1}$$

接下来我们主要讨论单模的情况,共振频率表示为 ω_{cav}。镜子的有限透射率、内部吸收以及散射到腔外等因素导致腔的有限衰减率,一般用 κ 表示。光学精细度 F 是一个非常有用的物理量,它给出了一个光子离开腔前在腔中往返传播的次数,与自由光谱范围的关系是 $F \equiv \frac{\Delta\omega_{\text{FSR}}}{\kappa}$。此外,光学腔的品质因数定义为 $Q_{\text{opt}} = \omega_{\text{cav}} \tau$,其中 $\tau = \kappa^{-1}$ 是光子的寿命。注意品质因数也可以用来表征机械振子的衰减率。一般来讲,腔的衰减率 κ 可以包含两方面的贡献,一个是与输入、输出耦合有关的损耗,另一个是内部损耗。对于品质因数很高的腔而言,腔总的衰减率可以写成上述两部分贡献之和,即 $\kappa_{\text{ex}} + \kappa_0$,其中 κ_{ex} 是与输入耦合有关的衰减率,而 κ_0 代表的是内部衰减率。对于法布里-珀罗腔而言,κ_{ex} 代表的是输入腔镜的衰减率,而 κ_0 表示腔内的衰减率,包括第二个腔镜的透射损耗以及第一个腔镜的散射和吸收损耗等。

耦合到外部电磁环境的光学腔的量子力学描述可以通过主方程或者输入-输出理论给出。腔场算符 a 的时间演化可以通过海森伯运动方程得到,其中 a 的衰减率为 $\kappa/2$。与此同时,腔场的涨落不断受到进入到腔的量子噪声的影响。在这里讨论的情况中,我们区分开输入耦合带来的外部衰减率 κ_{ex} 以及内部衰减率 κ_0。腔场的运动方程为[34]

$$\dot{a} = -\frac{\kappa}{2} a + \mathrm{i}\Delta a + \sqrt{\kappa_{\text{ex}}} a_{\text{in}} + \sqrt{\kappa_0} f_{\text{in}} \tag{1.2.2}$$

经典运动方程可以通过取平均值得到,即用 $\langle a \rangle$ 代替 a。上述方程已经变换到了激光频率 ω_L 的旋转框架下,即 $a^{\text{orig}} = \mathrm{e}^{-\mathrm{i}\omega_L t} a^{\text{here}}$,并且定义了激光与腔模之间的失谐量 $\Delta = \omega_L - \omega_{\text{cav}}$。对于机械振子而言,也存在着类似的运动方程来描述耗散以及相应的噪声。

输入场 $a_{\text{in}}(t)$ 应该被认为是随机的量子场。在最简单情况下,它包含 t 时刻耦合到腔的涨落真空电场和相干激光驱动场。同样的公式还可以用来描述压缩态或者其他更复杂的态。$P = \hbar \omega_L \langle a_{\text{in}}^+ a_{\text{in}} \rangle$ 是进入到腔的输入功率,即 $\langle a_{\text{in}}^+ a_{\text{in}} \rangle$ 是单位时间内到达腔的光子数。根据开放量子系统的输入-输出理论,从法布里-珀罗腔反射的场由下式给出:

$$a_{\text{out}} = a_{\text{in}} - \sqrt{\kappa_{\text{ex}}} a \tag{1.2.3}$$

这个输入-输出关系正确描述了从法布里-珀罗腔的输入镜反射的场。此外，它还可以描述耦合的单向波导-振子系统（比如回音壁模式的振子耦合于一个波导）的透射的泵浦场。

1.2.2 机械振子

法布里-珀罗腔右边可移动的镜子可看作频率为 ω_m 的机械振子，机械振子能量的损耗由能量衰减率 γ_m 表示，与机械振子品质因数之间的关系是 $Q_m = \omega_m/\gamma_m$。机械振子的位移 $x(t)$ 随时间演化的方程为

$$m_{\text{eff}} \frac{\mathrm{d}x^2(t)}{\mathrm{d}t^2} + m_{\text{eff}} \gamma_m \frac{\mathrm{d}x(t)}{\mathrm{d}t} + m_{\text{eff}} \omega_m^2 x(t) = F_{\text{ex}}(t) \qquad (1.2.4)$$

式中，m_{eff} 代表机械振子的有效质量，$F_{\text{ex}}(t)$ 代表作用在机械振子上所有力的和。在没有任何外力的情况下，$F_{\text{ex}}(t)$ 表示热的朗之万力。此外，这里假定了机械振子的衰减率 γ_m 与频率无关。方程(1.2.4)在频域空间中很容易求解。引入傅里叶变换 $x(\omega) = \int_{-\infty}^{+\infty} \mathrm{d}t\, e^{i\omega t} x(t)$，那么 $x(\omega) = \chi_m(\omega) F_{\text{ex}}(\omega)$ 定义了将外力和位置变化联系起来的极化率 χ_m，表达式为

$$\chi_m(\omega) = [m_{\text{eff}}(\omega_m^2 - \omega^2) - i m_{\text{eff}} \gamma_m \omega]^{-1} \qquad (1.2.5)$$

低频响应由 $\chi_m(0) = (m_{\text{eff}} \omega_m^2)^{-1} = 1/k$ 给出，其中 k 代表弹簧常数。

机械振子的哈密顿量为

$$H = \hbar \omega_m b^\dagger b + \frac{1}{2} \hbar \omega_m \qquad (1.2.6)$$

机械振子的零点能 $(1/2)\hbar\omega_m$ 可以忽略掉。$b^\dagger$ 和 b 分别表示机械振子的声子产生和湮灭算符，与位移和动量的关系为

$$x = x_{\text{zpf}}(b^\dagger + b), \quad p = -i m_{\text{eff}} \omega_m x_{\text{zpf}}(b - b^\dagger) \qquad (1.2.7)$$

式中，

$$x_{\text{zpf}} = \sqrt{\frac{\hbar}{2 m_{\text{eff}} \omega_m}} \qquad (1.2.8)$$

是机械振子的零点涨落振幅，即位移在基态中的期待值 $\langle 0|x^2|0\rangle = x_{\text{zpf}}^2$，其中 $|0\rangle$ 表示机械真空态。位置和动量满足对易关系 $[x,p] = i\hbar$。此外，$b^\dagger b$ 是声子数算符，平均值为 $\bar{n} = \langle b^\dagger b \rangle$。

1.2.3 光力耦合

法布里-珀罗腔中的光子入射到机械振子上后会被反射,光子的动量发生改变,从而对机械振子产生辐射压力。单个光子的动量转移为 $|\Delta p|=2h/\lambda$(λ 是光子的波长)。因此,辐射压力为

$$\langle F\rangle = 2\hbar k\frac{\langle a^\dagger a\rangle}{\tau_c}=\hbar\frac{\omega_{\text{cav}}}{L}\langle a^\dagger a\rangle \qquad (1.2.9)$$

其中,$\tau_c=2L/c$ 表示腔内光子的往返时间。因此,$\hbar\omega_{\text{cav}}/L$ 表示腔内一个光子产生的辐射压力。参数 ω_{cav}/L 描述腔共振频率随着位置的改变。一般情况下,光力耦合可以通过几种方式产生,包括反射时的直接动量转移(法布里-珀罗腔中有一个可移动的镜子),通过腔频的色散移动形成耦合(中间有一个薄膜、受限在腔中的悬浮纳米颗粒),或者通过光学近场效应(比如振子倏逝场中的纳米颗粒)等。

没有耦合的光学(ω_{cav})和机械(ω_m)模式可由两个简谐振子表示,哈密顿量为

$$H_0=\hbar\omega_{\text{cav}}a^\dagger a+\hbar\omega_m b^\dagger b \qquad (1.2.10)$$

当腔共振频率受到机械振子的振幅调制时,

$$\omega_{\text{cav}}(x)=\omega_{\text{cav}}+x\partial\omega_{\text{cav}}/\partial x+\cdots \qquad (1.2.11)$$

对大多数光力实验而言,上述方程保留线性项即可。定义光学腔频率随机械振子位移的变化率为 $G=-\partial\omega_{\text{cav}}/\partial x$。劳(Law)等给出了光力系统哈密顿量的更加详细的推导过程[35]。在长度为 L 的腔中,$G=\omega_{\text{cav}}/L$。如果 $x>0$ 表示腔的长度增加,会导致共振频率 $\omega_{\text{cav}}(x)$ 减小,从而 $G>0$。一般情况下,保留到位移的一阶项,可以得到

$$\hbar\omega_{\text{cav}}(x)a^\dagger a\approx\hbar(\omega_{\text{cav}}-Gx)a^\dagger a \qquad (1.2.12)$$

因此,腔和机械振子的相互作用哈密顿量可表示为

$$H_{\text{int}}=-\hbar g_0 a^\dagger a(b^\dagger+b) \qquad (1.2.13)$$

式中,

$$g_0=Gx_{\text{zpf}} \qquad (1.2.14)$$

是真空光力耦合强度,表示单个光子和单个声子之间的相互作用。相互作用哈密顿量揭示出腔与机械振子之间的相互作用本质上是一种非线性过程,包

含了三个算符(三波混频)。辐射压力是相互作用哈密顿量 H_{int} 对位移求导

$$F = -\frac{dH_{int}}{dx} = \hbar G a^\dagger a = \hbar \frac{g_0}{x_{zpf}} a^\dagger a \quad (1.2.15)$$

整个系统的哈密顿量 H 还包括描述耗散(光子衰减和机械摩擦)、涨落和外加激光的驱动等项。为了研究方便,通常将光场变换到激光频率 ω_L 的旋转框架下。利用幺正变换 $U = \exp(i\omega_L a^\dagger a t)$ 可以使驱动项变得与时间无关,并且得到一个新的哈密顿量

$$\begin{aligned} H &= U H_{old} U^\dagger - i\hbar U \partial U^\dagger / \partial t \\ &= -\hbar \Delta a^\dagger a + \hbar \omega_m b^\dagger b - \hbar g_0 a^\dagger a (b^\dagger + b) + \cdots \end{aligned} \quad (1.2.16)$$

其中 $\Delta = \omega_L - \omega_{cav}$ 是激光与腔场之间的失谐量,并且上述方程中没有包含驱动和耗散项。驱动项可写为 $H_{drive} = i\hbar\sqrt{\kappa_{ex}} a^\dagger \alpha_{in} + H.c.$,(Hermitian conjugation 厄米共轭)其中 α_{in} 代表激光的振幅。

接下来介绍腔光力学系统中所谓的线性化近似描述。为了进行线性化,腔场可分成平均相干振幅 $\langle a \rangle = \bar{\alpha}$ 和涨落项,即

$$a = \bar{\alpha} + \delta a \quad (1.2.17)$$

哈密顿量中相互作用部分

$$H_{int} = -\hbar g_0 (\bar{\alpha} + \delta a)^\dagger (\bar{\alpha} + \delta a)(b + b^\dagger) \quad (1.2.18)$$

可以根据 $\bar{\alpha}$ 的级次进行展开。第一项 $-\hbar g_0 |\bar{\alpha}|^2 (b+b^\dagger)$ 是 α 的平方项,代表平均辐射压力 $\bar{F} = \hbar G |\bar{\alpha}|^2$,可以通过先将位移原点作一个平移 $\delta\bar{x} = \bar{F}/m_{eff}\omega_m^2$,然后使用修正后的失谐量 $\Delta_{new} \equiv \Delta_{old} + G\delta\bar{x}$ 而忽略掉。第二项是 $\bar{\alpha}$ 的一次项,需要保留

$$-\hbar g_0 (\bar{\alpha}^* \delta a + \bar{\alpha} \delta a^\dagger)(b + b^\dagger) \quad (1.2.19)$$

第三项 $-\hbar g_0 \delta a^\dagger \delta a$ 是 α 的零次项,因为非常小而忽略掉。不失一般性,我们假定 $\sqrt{n_{cav}}$ 为实数。因此,旋转框架下的哈密顿量为

$$H \approx -\hbar \Delta \delta a^\dagger \delta a + \hbar \omega_m b^\dagger b + H_{int}^{(lin)} + \cdots \quad (1.2.20)$$

其中相互作用部分

$$H_{int}^{(lin)} = -\hbar g_0 \sqrt{n_{cav}} (\delta a^\dagger + \delta a)(b^\dagger + b) \quad (1.2.21)$$

成为线性化的,因为相应的耦合运动方程在这种近似下是线性的。$g = g_0 \sqrt{n_{cav}}$ 通常称为有效光力耦合强度,显然它依赖于激光的强度。线性化

描述即使在腔内光子数不是非常大时也比较好，原因是衰减率 κ 比较大时机械系统不能分辨单个的光子。当 $g > \kappa$ 时腔光力学系统进入强耦合区域，目前实验上已经可以达到强耦合区域。另一更难达到的区域是 $g_0 > \kappa$，即单光子耦合强度超过了腔的衰减率，在这一区域更容易观察到非线性量子效应。

在可分辨边带区域 ($\kappa < \omega_m$)，失谐量不同时相互作用式 (1.2.21) 可以分为三种不同的情况。当 $\Delta \approx -\omega_m$ (红失谐区域) 时，机械振子和驱动腔模之间可以互相交换量子，相互作用哈密顿量化简为

$$-\hbar g (\delta a^\dagger b + \delta a b^\dagger) \tag{1.2.22}$$

同时产生或湮灭两个量子的项 ($\delta a^\dagger b^\dagger$ 和 $\delta a b$)，因为非共振而被忽略掉。只保留方程 (1.2.22) 中的共振项也称为旋转波近似。红失谐驱动时可以实现机械振子的冷却以及光和机械振子之间的量子态转移。在量子光学领域，方程 (1.2.22) 通常称为分束器形式的哈密顿量。

当 $\Delta \approx \omega_m$ (蓝失谐区域) 时，旋转波近似后的保留项

$$-\hbar g (\delta a^\dagger b^\dagger + \delta a b) \tag{1.2.23}$$

代表双模压缩相互作用，可以用来实现参数放大[36]。没有耗散时，这种形式的相互作用可以导致存储在力学模式和驱动光学模式中的能量指数增长，并且机械和光学模式之间存在强的量子关联。因此，这种相互作用可以实现光学模式和力学模式之间的量子纠缠。就力学模式而言，能量的增加可被认为是反阻尼或者放大。如果机械振子的损耗足够低，这种行为可以激发动态不稳定进而导致自持续振荡。

当 $\Delta = 0$ 时，相互作用

$$-\hbar g (\delta a^\dagger + \delta a)(b + b^\dagger) \tag{1.2.24}$$

意味着机械振子的位移 $x \propto b + b^\dagger$，导致光场的相位移动，可以用于光力位移探测。此外，这种哈密顿量可以用于对光场振幅分量 $\delta a^\dagger + \delta a$ 进行量子非破坏性测量，原因是这个算符对整个哈密顿量对易。

1.3 几种典型的腔光力学系统

近年来，随着纳米科技的发展，实验上已能够制备出很多不同的腔光力学系统，如图 1.3.1 所示。下面简单介绍几种典型的腔光力学系统。

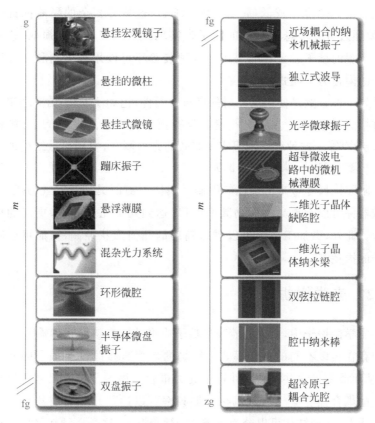

图 1.3.1 实验上实现的各种腔光力学系统[34]

（请扫Ⅱ页二维码看彩图）

1.3.1 悬挂的镜子

实现光力相互作用的一个非常明显的方法就是悬挂光学腔中的一个镜子。镜子的机械运动直接改变腔的长度，从而改变腔场的频率响应。实验上第一个实现这种类型的光力系统采用的是包含可移动镜子的法布里-珀罗腔，最早可以回溯到引力波探测中的激光干涉仪[37]。尽管悬挂宏观腔镜的目的是实现声学隔离，光力效应尤其是量子力学的辐射压涨落最终决定了干涉灵敏度的极限[38]。与此同时，这种结构使得人们可以利用腔光力学研究真正宏观的测试质量的质心运动。到目前为止，这种类型的光力实验演示了光学双稳、光学弹簧效应[39]、光学冷却[40]和反馈冷却[41]等。这些实验的一个困难是机械振动的频率比较低($\omega_m/2\pi < 1$ kHz),这需要复杂的隔离抵抗声学噪声,主要通过分几步悬挂宏观的镜子来实现。为了减小机械损耗,最近有人提出将宏观的镜子悬置在光阱中。这种装置还可以用来监测和光力调控宏观镜子的内在力学模式[17]。

另外一种方案是使用高反射的微机械器件,比如法布里-珀罗端镜。这些系统包括涂层悬臂[42]、微柱[43]、梁和悬臂顶部微米级的镜垫[44]、微机械悬浮光学涂层[45]等。这种结构中想实现有效的光力耦合,要求机械结构的尺寸比光的波长要大很多:典型的腔长为 $10^{-5} \sim 10^{-2}$ m,光学精细度达 10^5,这主要受到有限腔镜尺寸带来的损耗的限制。相比于前面提到的宏观的镜子,这些微机械器件可以有更高的振动频率(可以达到几十兆赫兹)和更高的机械品质因数。此外,微加工技术可以进行精确的几何形状控制,从而可以减小机械振子的损耗[46]。

1.3.2 光学微振子

在光学微振子中,光以沿着环形振子边缘的回音壁模式传输[47]。在这种结构中存在很多不同的力学模式。结构的形变直接改变了振子的光学路径长度,从而改变光学共振频率并产生光力耦合。微振子的小尺寸可以允许实现较大的耦合强度 g_0[48],且机械共振频率可以达到几兆赫兹甚至几千兆赫兹。本质上,这种结构可以分为三类:①微盘振子,这是一种在平面光子电路中标准的振子结构。最近实验中[49]的光力耦合强度可达 $g_0 \approx 2\pi \times 8 \times 10^5$ Hz。②微球振子。这种结构可以实现较大的光学品质因数[50],而机械品质因数主要受到内部材料损耗的限制,尤其是经常使用的硅微球。③微环形振子,这种结构可以由微盘振子通过热回流过程将侧壁熔化成环形拓扑得到。产生的光滑的表面以及微加工控制使得这种振子可以有较高的光学和机械品质因数[51],这也导致了第一个关于辐射压驱动的光力参数放大和可分辨边带操作的实验演示[52]。这种几何结构的优点是拥有好的光学品质和可分辨的边带区域,本质上是由于机械振子的共振频率可以从 10 MHz 到几千兆赫兹。

1.3.3 波导和光子晶体腔

片上波导和光子晶体腔是另外一种实现光力耦合的结构。光子晶体主要通过周期性地调制一些材料(比较典型的是硅)的折射率形成,在晶格中产生一些类似于电子能带的光带。光不能在带隙中传输。当一些人造的缺陷引入周期性结构中时,将形成一些局域的电磁场模式[53]。这种结构称为光子晶体腔。为了实现光力器件,面内光子晶体腔被钻蚀成纳米机械梁。机械运动导致腔边界和材料内的压力发生变化,从而引起结构中腔光子和力学模式的光力耦合。马尔多瓦(Maldova)和托马斯(Thomas)预言了光子晶体中同时出现局域的光学和振动缺陷模式[54]。光子晶体中的光力耦合在一维[55]和二维[56]光子晶体腔中都得到了实验证实。腔的尺寸比较小,再加上局域力学模式的质量很小导致该结构中的光力耦合强度比通常的法布里-珀罗腔中的要大很多,目前实验上可达 $g_0/(2\pi) \approx 1$ MHz。力学模式的频率可从几十兆赫兹到几千兆赫兹,这样可以大大减小环境的热占据数 $\bar{n}_{th} \approx k_B T/\hbar \omega_m$。利用周期性边界条件创建带隙的办法可以拓展到力学模式中。通过引入与声子波长相匹配的周期性结构可以产生包含局域化光子和声子模式的一维光子晶体腔,其中力学模式的品质因数可以被极大提高[57]。

目前,由于这种结构中的光力耦合强度比较大,系统可以进入非线性光子-声子相互作用区域。此外,千兆赫兹量级的机械共振频率使得低温操作时的平均声子数即使在没有附加激光冷却时仍然可以低于1,这对量子应用非常有利。最后,面内的结构可以与集成(硅)光子结构兼容,从而提供了一种实现更大尺寸的光机械阵列的途径,这在经典和量子信息处理以及集体动力学研究中都是非常重要的。

1.3.4 悬浮的纳米物体

这种类型的腔光力学结构是使用两端镜子都是固定的光学腔,此外,一种机械元件位于腔内或者在腔的近场中。这种结构中腔和亚波长尺寸的机械物体之间可以形成有效的光力耦合,在法布里-珀罗腔中悬挂高品质的机械薄膜[58]、氮化硅[59]、铝镓砷[60]、碳纳米线[61]等都属于这种结构。嵌入的纳米物体可以通过调制色散或者耗散来改变腔场[62]。

另外一种基于法布里-珀罗振子的方法是利用光学微振子表面的近场效应,其中倏逝光场可以形成与其他结构的色散耦合。本质上,机械运动调节不同界面之间的距离 d。因为近场效应,光力耦合强度随着距离 d 指数增加,所以可以产生较大的光力耦合强度 g_0。这种结构已经被用来产生环形微腔和附近的氮化硅纳米机械振子之间的光力耦合[63]，另外一种可能性是利用光学近场耦合两个机械振动的微盘振子[64]或两个光子晶体腔[65]。

为了进一步抑制机械损耗，可以将机械物体悬浮在附加的光学偶极子阱或者腔中的驻波场中[66]。当纳米物体被放置在腔模的驻波场中时，光力作用是平方耦合型的。当物体处于节点或者反节点时，光学频率移动不再是线性的，而是与机械振子位移的平方有关。这可以产生一些有趣的应用，包括单声子的量子非破坏性测量。这种装置还可以用来实现两个纳米物体（比如机械薄膜和单个原子）[67]之间的强耦合。

1.3.5　微波振子

与光学微腔类似，LC 电路形成一种微波领域的电磁辐射的振子，即 $\omega_c/(2\pi) \approx$ GHz。电容耦合于微波腔的机械振子的运动导致耦合电容及 LC 共振频率（$\partial C/\partial x \propto \partial \omega_c/\partial x$）的改变。因此，这是一种标准的腔光力学辐射压相互作用。这方面的早期实验由布拉金斯基于 1967 年[9]和马努金（Manukin）于 1977 年[11]完成，之后在引力波探测中也用到了这种结构[68]。这些工作证实了阻尼和光力反作用效应，比如冷却和参数放大。后来，离子阱实验中通过 LC 电路实现了微机械振子的冷却[69]。随着微加工超导电路的出现，人们成功地将纳米机械振子耦合于微波腔[70]。典型的机械振子的频率通常是几兆赫兹到几十兆赫兹，而微波光子需要保持在低温环境下。对于几千兆赫兹的光子，环境温度低到毫开尔文量级就足够，这可以通过低温制冷机实现。虽然微波光子的动量转移比光学频率的光子的动量转移小几个数量级，但是光力耦合强度 g_0 却可以做到差不多大。

此外，电容耦合还被用来将纳米机械振子直接耦合于超导库珀对盒子[71]或超导相位量子比特[72]等两能级量子系统。值得指出的是，电容耦合不是必需的。最近，微波振子通过电介质梯度力耦合于纳米梁的振动[73]，这就扩大了可用材料的范围，在实际应用中是非常有益的。

1.3.6　超冷原子

腔光力学系统还可以由包含 10^6 个原子的云构成，其中原子云的集体运动就像单个力学模式。对于超冷原子而言,它们已经被冷却到了量子基态。在一个实验中,法布里-珀罗腔中的超冷铷原子云的集体运动被用来观察散粒噪声辐射压涨落的特征[74]。原子云的集体运动与光学腔场之间的色散耦合导致位置依赖的频率移动,从而形成了光力相互作用。假设驻波场反节点处单个原子产生的单光子色散能量移动是 $\delta E = -\hbar (g_0^{at})^2/\Delta_{at}$,其中 g_0^{at} 是原子-腔真空拉比频率,Δ_{at} 是原子与腔共振之间的失谐量,那么腔模和位于 x 附近的 N 个原子构成的原子云之间的耦合哈密顿量为 $N\delta E a^+ a \sin^2[k(\bar{x}+x)]$。展开到 x 的最低阶可以得到光力耦合强度[75]

$$g_0 = \frac{(g_0^{at})^2}{\Delta_{at}}(kx_{zpf}^{atom})\sin(2k\bar{x})\sqrt{N} \qquad (1.3.1)$$

式中,$x_{zpf}^{atom} = \sqrt{\hbar/2m_{atom}\omega_m}$,表示势阱中单个原子的零点涨落。注意 x_{zpf}^{atom} 比原子云质心运动的 $x_{zpf}^{c.m.}$ 大了 $\sqrt{N}$ 倍,这里我们假定了原子云的尺度比波长的小。当原子被束缚在节点或者反节点处时,最主要的光力耦合是 x^2 而不是 x。

另外一个实验中,腔光力学被用来冷却光学腔中热的铯原子云的运动[76]。此外,包含 10^6 个原子的玻色-爱因斯坦凝聚体的密度涨落被用来作为法布里-珀罗腔中的力学模式。在超冷原子情况下,由于强的色散耦合和小的质量(导致大的零点运动振幅),系统接近于单光子强耦合区域 $g_0/\kappa > 1$。

1.4 腔光力学系统的光学响应

光学腔与机械振子之间的光力耦合一方面可以用于实现机械振子的量子基态冷却和量子纠缠等,另一方面也会导致系统的光学响应发生变化,产生光力诱导透明、光力诱导吸收和放大等现象。

1.4.1 光力诱导透明

电磁诱导透明(electromagnetically induced transparency, EIT)[77]是一种发生在多能级原子中的现象,指的是在辅助激光场作用下原子的吸收率可变为零,主要是由干涉效应或者激发态中的暗态共振引起的。电磁诱导透明现象已经在冷原子以及其他固体系统中被观察到并有着一些重要的应用,比如光脉冲存储、慢光或者快光效应等。在腔光力学系统中也可以出现类似的现象，称为光力诱导透明（optomechanically induced transparency，OMIT）。最早是阿加瓦尔（Agarwal）和黄（Huang）[78]等理论上进行了预言，后来被魏斯（Weis）等[79]和萨法维-奈尼（Safavi-Naeini）等[80]分别在实验上观察到，如图1.4.1所示。当光力系统受到一束红失谐的强控制激光驱动时，产生的反斯托克斯场与近共振的弱探测场之间的量子干涉可以导致探测场的透射率接近于1。一束强的控制激光（$\bar{s}e^{-i\omega_c t}$）和一束弱的探测激光（$\delta s e^{-i\omega_p t}$）同时出现时，导致弱探测激光的透射率为

$$|t_p|^2 = \left|1 - \eta\kappa \frac{\chi_{\text{opt}}(\Omega)}{1 + g_0^2 \bar{a}^2 \chi_{\text{mech}}(\Omega)\chi_{\text{opt}}(\Omega)}\right|^2 \quad (1.4.1)$$

式中，$\Delta = \omega_c - \omega_{\text{cav}}$代表强控制场和腔共振频率$\omega_{\text{cav}}$之间的失谐量，$\Omega$代表控

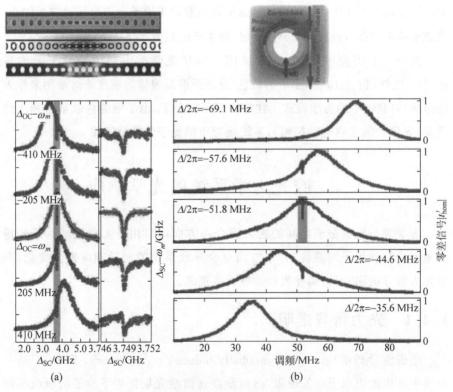

图1.4.1 实验上观察到的光力诱导透明现象[79,80]

（请扫Ⅱ页二维码看彩图）

制激光和探测激光之间的失谐量,即 $\Omega = \omega_p - \omega_c$。此外,这里还引入了耦合效率 $\eta = \kappa_{ex}/\kappa$,机械极化率 $\chi_{mech}^{-1}(\Omega) = -i(\Omega - \omega_m) + \gamma_m/2$,以及光学极化率 $\chi_{opt}^{-1}(\Omega) = -i(\Omega - \omega_m) + \kappa/2$。根据方程(1.4.1)画图后可以发现,当满足共振条件 $\Omega = \omega_m$ 时,透射谱中会出现透明窗口。当调节控制激光满足 $\Delta = -\omega_m$ 时,在 $\Omega = \omega_m$ 附近探测场透射谱的表达式可化简为

$$|t_p|^2 = \left| \frac{4\bar{n}_{cav}g_0^2}{4\bar{n}_{cav}g_0^2 + \gamma_m\kappa - 2i(\Omega - \omega_m)\kappa} \right| \tag{1.4.2}$$

这里为了简单起见,我们假设了临界耦合条件,即 $\eta = 1/2$。当 $\Omega = \omega_m$ 时,

$$|t_p|^2 = \left(\frac{C}{C+1} \right)^2 \tag{1.4.3}$$

这里 C 代表光力协同性,定义为 $C = 4g_0^2 \bar{n}_{cav}/\kappa\gamma_m$。

产生透明窗口的物理原因可以通过探测场和控制场之间的拍频产生的随时间变化的辐射压力进行解释。如果拍频正好等于机械振动的频率,机械振子将相干振动起来,并进一步产生斯托克斯场和反斯托克斯场。在可分辨边带极限下,斯托克斯场和腔场之间是非共振的,因此可以被忽略掉。但是反斯托克斯场与探测场之间是共振的,它们之间会发生干涉并导致透明窗口。因此,这种现象主要是由产生的反斯托克斯光子和探测光子之间的相消干涉引起。在弱耦合区域,透明窗口的宽度由总的有效机械衰减率给出:

$$\Gamma_{OMIT} = \gamma_m + 4g_0^2 \bar{n}_{cav}/\kappa = \gamma_{eff} \tag{1.4.4}$$

窄的透明窗口伴随着透射探测场快速的相位色散,这意味着带宽小于 Γ_{OMIT} 的脉冲将经历群速度延迟且脉冲不会发生失真。$\Delta = -\omega_m$ 时透射探测场的相位变化由下式给出:

$$\phi = \arctan\left(\frac{2(\Omega - \omega_m)\kappa}{4\bar{n}_{cav}g_0^2 + \gamma_m\kappa} \right) \tag{1.4.5}$$

失谐量为零且 $\omega = \omega_m$ 时,从上式可以得到群速度延迟为

$$\tau_g = \frac{d\phi}{d\Omega} = \frac{1}{\gamma_m} \frac{2}{C+1} = \frac{2}{\Gamma_{OMIT}} \tag{1.4.6}$$

但是,为了保持脉冲传播时不发生失真,脉冲的带宽需要比透明窗口的宽度小,这就限制了延迟-带宽积为 $\tau_g \Gamma_{OMIT} \approx 2$。使用腔光力学系统阵列可以增加延迟-带宽积,因此提供了一种将光信号存储在寿命较长的声子中的方法。这种级联的光力系统可以通过光子晶体实现[81]。

1.4.2 光力诱导吸收和放大

当控制激光作用在光力系统的蓝边带时($\Delta=\omega_m$),透射的探测场可以被放大[82,83]。在原子物理中类似的效应称为电磁诱导吸收(electromagnetically induced absorption,EIA)[84],如图 1.4.2 所示。产生的斯托克斯场和探测场之间的相长干涉可以放大弱的探测信号,与上述光力诱导透明现象类似。理论上这种现象也可以通过与光力诱导透明现象相同的方程进行描述,不同的是随着蓝失谐的控制场功率的增加,机械振子的有效衰减率减小。最大增益由蓝失谐的控制场的最大功率决定,而最大功率受到参数振荡不稳定性的限制。这种情况下最大的平均增益由公式 $G_{av}(\Delta=\omega_m)\approx 4(4\omega_m/\kappa)^2$ 给出。对于任何非简并的参数放大器而言,这种放大过程一定会增加至少 1/2 个噪声量子数[36]。存在热涨落时,总的增加的噪声 $\bar{n}_{add}=\bar{n}_m+1/2\approx k_B T/\hbar\omega_m$。需要指出的是,这种光学放大过程不会像光学布里渊散射一样产生受激光学放大过程,原因是这里的光学放大过程中光学耗散比机械耗散要大很多($\kappa\gg\gamma_m$)。

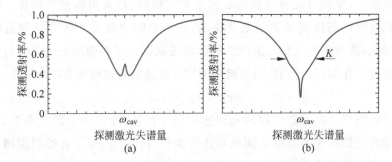

图 1.4.2 红边带的控制场(a)和蓝边带的控制场作用下探测场的透射谱[34](b)

(请扫 II 页二维码看彩图)

参 考 文 献

[1] LEBEDEW P. Untersuchungen über die druckkräfte des lichtes[J]. Ann. Phys., 1901, 311: 433.

[2] EINSTEIN A. On the development of our views concerning the nature and constitution of radiation[J]. Z. Phys., 1909, 10: 817.

[3] FRISCH O. Experimenteller nachweis des einsteinschen strahlungsrückstoßes[J].

Z. Phys. B, 1933, 86: 42.

[4] BETH R. Mechanical detection and measurement of the angular momentum of light [J]. Phys. Rev. , 1936, 50: 115.

[5] ASHKIN A. Trapping of atoms by resonance radiation pressure[J]. Phys. Rev. Lett. , 1978, 40: 729.

[6] HÄNSCH T W, SCHAWLOW A L. Cooling of gases by laser radiation[J]. Opt. Commun. , 1975, 13: 68.

[7] WINELAND D J, DEHMELT H. Proposed 1014 delta upsilon less than upsilon laser fluorescence spectroscopy on t1+ mono-ion oscillator[J]. Bull. Am. Phys. Soc. , 1975, 20: 637.

[8] BLOCH I, DALIBARD J, ZWERGER W. Many-body physics with ultracold gases [J]. Rev. Mod. Phys. , 2008, 80: 885.

[9] BRAGINSKY V B, MANUKIN A B. Ponderomotive effects of electromagnetic radiation[J]. Sov. Phys. JETP, 1967, 25: 653.

[10] DORSEL A, MCCULLEN J D, MEYSTRE P, et al. Optical bistability and mirror confinement induced by radiation pressure[J]. Phys. Rev. Lett. , 1983, 51: 1550.

[11] BRAGINSKY V B, MANUKIN A B. Measurement of weak forces in physics experiments[M]. Chicago: University of Chicago Press, 1977.

[12] CAVES C M. Quantum-mechanical radiation-pressure fluctuations in an interferometer [J]. Phys. Rev. Lett. , 1980, 45: 75.

[13] FABRE C, PINARD M, BOURZEIX S, et al. Quantum-noise reduction using a cavity with a movable mirror[J]. Phys. Rev. A, 1994, 49: 1337.

[14] JACOBS K, TOMBESI P, COLLETT M, et al. Quantum-nondemolition measurement of photon number using radiation pressure[J]. Phys. Rev. A , 1994, 49: 1961.

[15] MANCINI S, VITALI D, TOMBESI P. Optomechanical cooling of a macroscopic oscillator by homodyne feedback[J]. Phys. Rev. Lett. , 1998, 80: 688.

[16] HECHENBLAIKNER G, GANGL M, HORAK P, et al. Cooling an atom in a weakly driven high-Q cavity[J]. Phys. Rev. A, 1998, 58: 3030.

[17] COHADON P F, HEIDMANN A, PINARD M. Cooling of a mirror by radiation pressure[J]. Phys. Rev. Lett. , 1999, 83: 3174.

[18] TITTONEN I, BREITENBACH G, KALKBRENNER T, et al. Interferometric measurements of the position of a macroscopic body: towards observation of quantum limits[J]. Phys. Rev. A, 1999, 59: 1038.

[19] VOGEL M, MOOSER C, KARRAI K, et al. Optically tunable mechanics of microlevers[J]. Appl. Phys. Lett. , 2003, 83: 1337.

[20] MERTZ J, MARTI O, MLYNEK J. Regulation of a microcantilever response by force feedback[J]. Appl. Phys. Lett. , 1993, 62: 2344.

[21] ZALALUTDINOV M, ZEHNDER A, OLKHOVETS A, et al. Autoparametric

optical drive for micromechanical oscillators[J]. Appl. Phys. Lett., 2001, 79: 695.

[22] HÖHBERGER M, KARRAI C K. Cavity cooling of a microlever[J]. Nature, 2004, 432: 1002.

[23] CARMON T, ROKHSARI H, YANG L. Temporal behavior of radiation-pressure-induced vibrations of an optical microcavity phonon mode[J]. Phys. Rev. Lett., 2005, 94: 223902.

[24] ARCIZET O, COHADON P F, BRIANT T, et al. Radiation-pressure cooling and optomechanical instability of a micromirror[J]. Nature, 2006, 444: 71.

[25] GIGAN S, BÖHM H R, PATERNOSTRO M, et al. Self-cooling of a micromirror by radiation pressure[J]. Nature, 2006, 444: 67.

[26] SCHLIESSER A, DEL'HAYE P, NOOSHI N, et al. Radiation pressure cooling of a micromechanical oscillator using dynamical backaction[J]. Phys. Rev. Lett., 2006, 97: 243905.

[27] BRENNECKE F, RITTER S, DONNER T, et al. Cavity optomechanics with a Bose-Einstein condensate[J]. Science, 2008, 322: 235.

[28] SHELBY R M, LEVENSON M D, BAYER P W. Guided acoustic-wave Brillouin scattering[J]. Phys. Rev. B, 1985, 31: 5244.

[29] LI M, PERNICE W H P, XIONG C, et al. Harnessing optical forces in integrated photonic circuits[J]. Nature, 2008, 456: 480.

[30] REGAL C A, TEUFEL J D, LEHNERT K W. Measuring nanomechanical motion with a microwave cavity interferometer[J]. Nat. Phys., 2008, 4: 555.

[31] TEUFEL J D, LI D, ALLMAN M S, et al. Circuit cavity electromechanics in the strong-coupling regime[J]. Nature, 2011, 471: 204.

[32] LAHAYE M D, BUU O, CAMAROTA B, et al. Approaching the quantum limit of a nanomechanical resonator[J]. Science, 2004, 304: 74.

[33] CLELAND A N, ALDRIDGE J S, DRISCOLL D C, et al. Nanomechanical displacement sensing using a quantum point contact[J]. Appl. Phys. Lett., 2002, 81: 1699.

[34] ASPELMEYER M, KIPPENBERG T J, MARQUARDT F. Cavity optomechanics [J]. Rev. Mod. Phys., 2014, 86: 1391.

[35] LAW C K. Interaction between a moving mirror and radiation pressure: a Hamiltonian formulation[J]. Phys. Rev. A, 1995, 51: 2537.

[36] CLERK A A, DEVORET M H, GIRVIN S M, et al. Introduction to quantum noise, measurement, and amplification[J]. Rev. Mod. Phys., 2010, 82: 1155.

[37] ABRAMOVICI A, ALTHOUSE W E, DREVER R W P, et al. LIGO: the laser interferometer gravitational-wave observatory[J]. Science, 1992, 256: 325.

[38] CAVES C M. Quantum-mechanical radiation-pressure fluctuations in an interferometer[J]. Phys. Rev. Lett., 1980, 45: 75.

[39] SHEARD B, GRAY M, MOW-LOWRY C, et al. Observation and characterization of an optical spring[J]. Phys. Rev. A, 2004, 69: 051801.

[40] CORBITT T, WIPF C, BODIYA T, et al. Optical dilution and feedback cooling of a gram-scale oscillator to 6.9 mK[J]. Phys. Rev. Lett., 2007,99: 160801.

[41] ABBOTT B, ABBOTT R, ADHIKARI R, et al. Observation of a kilogram-scale oscillator near its quantum ground state[J]. New J. Phys., 2009, 11: 073032.

[42] ARCIZET O, COHADON P F, BRIANT T, et al. Radiation-pressure cooling and optomechanical instability of a micromirror[J]. Nature, 2006,444: 71.

[43] VERLOT P,TAVERNARAKIS A, MOLINELL, C et al. Towards the experimental demonstration of quantum radiation pressure noise [J]. C. R. Phys., 2011, 12: 826.

[44] KLECKNER D, W MARSHALL, DE DOOD J M A, et al. High finesse optomechanical cavity with a movable thirty-micron-size mirror[J]. Phys. Rev. Lett., 2006, 96: 173901.

[45] BÖHM H R, GIGAN S, LANGER G, et al. High reflectivity high-Q micromechanical Bragg mirror[J]. Appl. Phys. Lett., 2006, 89: 223101.

[46] ANETSBERGER G, RIVIÈRE R, SCHLIESSER A, et al. Ultralow-dissipation optomechanical resonators on a chip[J]. Nat. Photonics, 2008, 2: 627.

[47] VAHALA K J. Optical microcavities[J]. Nature, 2003, 424: 839.

[48] VERHAGEN E, DELÉGLISE S, WEIS S A, et al. Quantum-coherent coupling of a mechanical oscillator to an optical cavity mode[J]. Nature,2012,482: 63.

[49] DING L, BAKER C, SENELLART P, et al. Wavelength-sized GaAs optomechanical resonators with gigahertz frequency[J]. Appl. Phys. Lett., 2011, 98: 113108.

[50] PARK Y S, WANG H. Resolved-sideband and cryogenic cooling of an optomechanical resonator[J]. Nat. Phys., 2009, 5: 489.

[51] ARMANI D K, KIPPENBERG T J, SPILLANE S M, et al. Ultra-high-Q toroid microcavity on a chip[J]. Nature, 2003, 421: 925.

[52] KIPPENBERG T J, ROKHSARI H, CARMON T, et al. Analysis of radiation-pressure induced mechanical oscillation of an optical microcavity[J]. Phys. Rev. Lett., 2005, 95: 033901.

[53] VAHALA K. Optical microcavities[M]. Singapore: World Scientific, 2004.

[54] MALDOVA M, THOMAS E L. Simultaneous localization of photons and phonons in two-dimensional periodic structures[J]. Appl. Phys. Lett., 2006, 88: 251907.

[55] EICHENFIELD M, CAMACHO R, CHAN J, et al. A picogram- and nanometre-scale photonic-crystal optomechanical cavity[J]. Nature, 2009, 459: 550.

[56] SAFAVI-NAEINI A H, ALEGRE T P M, WINGER M, et al. Optomechanics in an ultra-high-Q two-dimensional photonic crystal cavity[J]. Appl. Phys. Lett., 2010, 97: 181106.

[57] CHAN J, ALEGRE T P M, SAFAVI-NAEINI A H, et al. Laser cooling of a

nanomechanical oscillator into its quantum ground state[J]. Nature, 2011, 478: 89.

[58] THOMPSON J D, ZWICKL B M, JAYICH A M, et al. Strong dispersive coupling of a high-finesse cavity to a micromechanical membrane[J]. Nature, 2008, 452: 72.

[59] WILSON D, REGAL C, PAPP S, et al. Cavity optomechanics with stoichiometric SiN films[J]. Phys. Rev. Lett., 2009, 103: 207204.

[60] LIU J, USAMI K, NAESBY A, et al. High-Q optomechanical GaAs nanomembranes[J]. Appl. Phys. Lett., 2011, 99: 243102.

[61] FAVERO I, STAPFNER S, HUNGER D, et al. Fluctuating nanomechanical system in a high finesse optical microcavity[J]. Opt. Express, 2009, 17: 12813.

[62] XUEREB A, SCHNABEL R, HAMMERER K. Dissipative optomechanics in a Michelson-Sagnac interferometer[J]. Phys. Rev. Lett., 2011, 107: 213604.

[63] ANETSBERGER G, ARCIZET O, UNTERREITHMEIER Q P, et al. Near-field cavity optomechanics with nanomechanical oscillators[J]. Nat. Phys., 2009, 5: 909.

[64] JIANG X, LIN Q, ROSENBERG J, et al. High-Q double-disk microcavities for cavity optomechanics[J]. Opt. Express, 2009, 17: 20911.

[65] ROH Y G, TANABE T, SHINYA A, et al. Strong optomechanical interaction in a bilayer photonic crystal[J]. Phys. Rev. B, 2010, 81: 121101(R).

[66] BARKER P F, SHNEIDER M N. Cavity cooling of an optically trapped nanoparticle[J]. Phys. Rev. A, 2010, 81: 023826.

[67] HAMMERER K, WALLQUIST M, GENES C, et al. Strong coupling of a mechanical oscillator and a single atom[J]. Phys. Rev. Lett., 2009, 103: 063005.

[68] BLAIR D G, IVANOV E N, TOBAR M E, et al. High sensitivity gravitational wave antenna with parametric transducer readout[J]. Phys. Rev. Lett., 1995, 74: 1908.

[69] BROWN K R, BRITTON J, EPSTEIN R J, et al. Passive cooling of a micromechanical oscillator with a resonant electric circuit[J]. Phys. Rev. Lett., 2007, 99: 137205.

[70] REGAL C A, TEUFEL J D, LEHNERT K W. Measuring nanomechanical motion with a microwave cavity interferometer[J]. Nat. Phys., 2008, 4: 555.

[71] LAHAYE M D, SUH J, ECHTERNACH P M, et al. Nanomechanical measurements of a superconducting qubit[J]. Nature, 2009, 459: 960.

[72] O'Connell A D, HOFHEINZ M, ANSMANN M, et al. Quantum ground state and single-phonon control of a mechanical resonator[J]. Nature, 2010, 464: 697.

[73] FAUST T, KRENN P, MANUS S, et al. Microwave cavity-enhanced transduction for plug and play nanomechanics at room temperature[J]. Nat. Commun., 2012, 3: 728.

[74] MURCH K W, MOORE K L, GUPTA S, et al. Observation of quantum-

measurement backaction with an ultracold atomic gas[J]. Nat. Phys., 2008, 4: 561.

[75] STAMPER-KURN D M. Cavity optomechanics[M]. New York: Springer, 2014.

[76] SCHLEIER-SMITH M H, LEROUX I D, ZHANG H, et al. Optomechanical cavity cooling of an atomic ensemble[J]. Phys. Rev. Lett., 2011, 107: 143005.

[77] FLEISCHHAUER M, IMAMOGLU A, MARANGOS J P. Electromagnetically induced transparency: optics in coherent media[J]. Rev. Mod. Phys., 2005, 77: 633.

[78] AGARWAL G S, HUANG S. Electromagnetically induced transparency in mechanical effects of light[J]. Phys. Rev. A, 2010, 81: 041803.

[79] WEIS S, RIVIERE R, DELEGLISE S, et al. Optomechanically induced transparency[J]. Science, 2010, 330: 1520.

[80] SAFAVI-NAEINI A H, ALEGRE T P M, CHAN J, et al. Electromagnetically induced transparency and slow light with optomechanics[J]. Nature, 2011, 472: 69.

[81] CHANG D E, SAFAVI-NAEINI A H, HAFEZI M, et al. Slowing and stopping light using an optomechanical crystal array[J]. New J. Phys., 2011, 13: 023003.

[82] MASSEL F, HEIKKILÄ T T, PIRKKALAINEN J M, et al. Microwave amplification with nanomechanical resonators[J]. Nature, 2011, 480: 351.

[83] HOCKE F, ZHOU X, SCHLIESSER A, et al. Electromechanically induced absorption in a circuit nano-electromechanical system[J]. New J. Phys., 2012, 14: 123037.

[84] LEZAMA A, BARREIRO S, AKULSHIN A. Electromagnetically induced absorption[J]. Phys. Rev. A, 1999, 59: 4732.

第 2 章 腔光力学系统中的慢光效应

2.1 双腔光力系统中的光力诱导透明和慢光效应

2.1.1 引言

新兴的腔光力学领域[1-4]研究通过辐射压实现光学模与力学模之间的相互作用,借此可以观察宏观系统的量子力学行为。有关制备和冷却技术的最近进展,为光力系统中实现单光子强耦合[5-10]及冷却纳米机械振子到量子基态铺平了道路[11,12]。此外,由于机械振子引起的相互作用改变了光机系统的光学响应特性,导致了正则模式分裂[13,14]和电磁诱导透明现象[15-18]。在电磁诱导透明现象中,不透明的介质可以在强泵浦光的作用下变得透明[19];伴随折射率的剧烈变化引起探测光的群速急剧减小,因而可以用于慢光甚至停光效应[20,21]。首次观察到电磁诱导透明现象是在原子蒸气中[22],最近在量子阱[23]、超材料[24]和氮空位中心[25]等各种固体系统中也观察到了电磁诱导透明现象。对于腔光力学系统,在光学领域[17,18]和微波领域[26]都已经观察到基于电磁诱导透明的慢光与快光现象。最近,在由超导微波谐振腔和纳米机械振子组成的微波光力系统中,当微波腔受到蓝失谐泵浦场作用时出现了光力诱导放大[27]和吸收[28]现象。

此外,两光学模耦合到同一个力学模的双腔光力系统已引起人们的广泛研究兴趣。杜布林德(Dobrind)等[29]业已证明,对于高频谐振子而言,双腔传感器能够导致达到标准量子极限(SQL)的功率显著减少。路德维希(Ludwig)等[30]和科尔马(Komar)等[31]从理论上证明了在双腔光机系统中量子非线性可以显著增强,这些结论可以用于光子和声子的光机量子信息处理之中[32]。然而,杜布林德(Dobrind)[29]、路德维希[30]和科尔马[31]的理论工作要求机械频率几乎是与光学能级分裂共振的,这对实验上的实现提出了更高要求。最近,曲(Qu)和阿加瓦尔(Agarwal)[33]在理论上证明,双腔光力系统可以用作存储元件以及光场变换。希尔(Hill)等[34]和董(Dong)等[35]也在实验上证明,在光力晶体纳米腔以及石英谐振腔中不同光波长的光子之间可实现相干波长转换。这些研究同时存在两个强泵浦光和一个弱探测光时双腔光力系统的光学响应特性。当把两个光学腔分别泵浦到它们的红边带时(即机械频率 ω_m 低于腔共振频率 ω_1 和 ω_2),在所探测的透射谱中将出现一个透明窗口。

2.1.2 模型和理论

考虑如图 2.1.1 所示的双腔光力系统,其中两个腔模 $a_k(k=1,2)$ 耦合到同一个力学模式 b。左腔受到频率为 ω_L、强度为 E_L 的强泵浦光及频率为 ω_p、强度为 E_p 的弱探测光同时驱动,而右腔则只受到频率为 ω_R、强度为 E_R 的强泵浦光驱动。在泵浦频率 ω_L 和 ω_R 的旋转框架下,双腔光力系统的哈密顿量可写为[34]

$$H = \sum_{k=1,2} \hbar \Delta_k a_k^\dagger a_k + \hbar \omega_m b^\dagger b - \sum_{k=1,2} \hbar g_k a_k^\dagger a_k (b^\dagger + b) + $$
$$i\hbar \sqrt{\kappa_{e,1}} E_L (a_1^\dagger - a_1) + i\hbar \sqrt{\kappa_{e,2}} E_R (a_2^\dagger - a_2) + $$
$$i\hbar \sqrt{\kappa_{e,1}} E_p (a_1^\dagger e^{-i\delta t} - a_1 e^{i\delta t}) \qquad (2.1.1)$$

式中,第一项描述共振频率为 $\omega_k(k=1,2)$ 的两个腔模能量,$a_k^\dagger(a_k)$ 是每个腔模的产生(湮灭)算符,$\Delta_1=\omega_1-\omega_L$ 和 $\Delta_2=\omega_2-\omega_R$ 是相应的腔场与泵浦场之间的失谐量。第二项表示产生(湮灭)算符为 $b^\dagger(b)$、共振频率为 ω_m 及有效质量为 m 的机械振子能量。第三项代表耦合度为 $g_k=(\omega_k/L_k)\sqrt{\hbar/(2m\omega_m)}$ 的辐射压相互作用,其中 L_k 是依赖于腔几何形状的有效长度。最后三项代表输入场,其中 E_L、E_R 和 E_p 取决于所用激光场的功率 P,关系分别为 $|E_L|=\sqrt{2P_L\kappa_1/\hbar \omega_L}$、$|E_R|=\sqrt{2P_R\kappa_2/\hbar \omega_R}$ 和 $|E_p|=\sqrt{2P_p\kappa_1/\hbar \omega_p}$($\kappa_k$ 为第 k 个腔模的线宽)。如同文献[34]真实双腔光力系统所提到的,腔模的总线宽 $\kappa_k=\kappa_{i,k}+\kappa_{e,k}$,$\kappa_{e,k}$ 是起因于耦合到外部光子波导而出现的腔衰减率。$\delta=\omega_p-\omega_L$ 是探测场与左泵浦场之间的失谐量。

对于 a_1、a_2 及定义为 $Q=b^\dagger+b$ 的算符,应用海森伯运动方程,并引入相

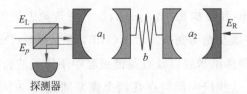

图 2.1.1 双腔光力系统示意图。其中两个腔模 a_1 和 a_2 耦合到同一个力学模式 b。左腔用强泵浦光 E_L 驱动,同时加上一弱探测光 E_p,而右腔只用泵浦光 E_R 驱动

(请扫Ⅱ页二维码看彩图)

应的阻尼和噪声项[36],得到量子朗之万方程如下:

$$\dot{a}_1 = -\mathrm{i}(\Delta_1 - g_1 Q)a_1 - \kappa_1 a_1 + \sqrt{\kappa_{e,1}}(E_L + E_p \mathrm{e}^{-\mathrm{i}\delta t}) + \sqrt{2\kappa_1}\, a_{in,1} \quad (2.1.2)$$

$$\dot{a}_2 = -\mathrm{i}(\Delta_2 - g_2 Q)a_2 - \kappa_2 a_2 + \sqrt{\kappa_{e,2}}\, E_R + \sqrt{2\kappa_2}\, a_{in,2} \quad (2.1.3)$$

$$\ddot{Q} + \gamma_m \dot{Q} + \omega_m^2 Q = 2g_1 \omega_m a_1^\dagger a_1 + 2g_2 \omega_m a_2^\dagger a_2 + \xi \quad (2.1.4)$$

式中,$a_{in,1}$ 和 $a_{in,2}$ 是均值为零的输入真空噪声算符,ξ 是零均值的布朗随机力[36]。

按照量子光学常规方法,通过令所有的时间导数为零,可以导出式(2.1.2)～式(2.1.4)的稳态解为

$$a_{s,1} = \frac{\sqrt{\kappa_{e,1}}\, E_L}{\kappa_1 + \mathrm{i}\Delta_1'},\quad a_{s,2} = \frac{\sqrt{\kappa_{e,2}}\, E_R}{\kappa_2 + \mathrm{i}\Delta_2'},\quad Q_s = \frac{2}{\omega_m}(g_1 |a_{s,1}|^2 + g_2 |a_{s,2}|^2)$$

(2.1.5)

式中,$\Delta_1' = \Delta_1 - g_1 Q_s$ 和 $\Delta_2' = \Delta_2 - g_2 Q_s$ 是包括辐射压效应在内的有效腔失谐量。可以将每个算符写成其稳态平均值及均值为零的涨落项的和的形式,即

$$a_1 = a_{s,1} + \delta a_1,\quad a_2 = a_{s,2} + \delta a_2,\quad Q = Q_s + \delta Q \quad (2.1.6)$$

将这些方程代入朗之万方程(2.1.2)～方程(2.1.4),并且假设 $|a_{s,1}| \gg 1$ 和 $|a_{s,2}| \gg 1$,可以忽略非线性项 $\delta a_1^\dagger \delta a_1$、$\delta a_2^\dagger \delta a_2$、$\delta a_1 \delta Q$ 及 $\delta a_2 \delta Q$。由于驱动场是弱的经典相干场,我们将所有算符用其期待值表示,并丢掉量子和热噪声项[36]。因而可以写出线性化朗之万方程:

$$\langle \delta \dot{a}_1 \rangle = -(\kappa_1 + \mathrm{i}\Delta_1)\langle \delta a_1 \rangle + \mathrm{i}g_1 Q_s \langle \delta a_1 \rangle + \mathrm{i}g_1 a_{s,1}\langle \delta Q \rangle + \sqrt{\kappa_{e,1}}\, E_p \mathrm{e}^{-\mathrm{i}\delta t}$$

(2.1.7)

$$\langle \delta \dot{a}_2 \rangle = -(\kappa_2 + \mathrm{i}\Delta_2)\langle \delta a_2 \rangle + \mathrm{i}g_2 Q_s \langle \delta a_2 \rangle + \mathrm{i}g_2 a_{s,2}\langle \delta Q \rangle \quad (2.1.8)$$

$$\langle\delta\ddot{Q}\rangle + \gamma_m \langle\delta\dot{Q}\rangle + \omega_m^2 \langle\delta Q\rangle = 2\omega_m g_1 a_{s,1}(\langle\delta a_1\rangle + \langle\delta a_1^+\rangle) +$$
$$2\omega_m g_2 a_{s,2}(\langle\delta a_2\rangle + \langle\delta a_2^\dagger\rangle) \quad (2.1.9)$$

为了求解式(2.1.7)～式(2.1.9)，作如下代换[37]：$\langle\delta a_1\rangle = a_{1+}e^{-i\delta t} + a_{1-}e^{i\delta t}$，$\langle\delta a_2\rangle = a_{2+}e^{-i\delta t} + a_{2-}e^{i\delta t}$，$\langle\delta Q\rangle = Q_+ e^{-i\delta t} + Q_- e^{i\delta t}$，分别代入式(2.1.7)～式(2.1.9)，求出以下解：

$$a_{1+} = \frac{\sqrt{\kappa_{e,1}} E_p}{\kappa_1 + i\Delta_1' - i\delta} - \frac{1}{d(\delta)} \frac{ig_1^2 n_1 \sqrt{\kappa_{e,1}} E_p}{(\kappa_1 + i\Delta_1' - i\delta)^2} \quad (2.1.10)$$

式中，

$$d(\delta) = \sum_{k=1,2} \frac{2\Delta_k' g_k^2 n_k}{(\kappa_k - i\delta)^2 + \Delta_k'^2} - \frac{\omega_m^2 - \delta^2 - i\delta\gamma_m}{\omega_m} \quad (2.1.11)$$

并且 $n_k = |a_{s,k}|^2$。该 n_k 约等于每个腔内的泵浦光子数，由下述耦合方程确定：

$$n_1 = \frac{\kappa_{e,1} E_L^2}{\kappa_1^2 + [\Delta_1 - 2g_1/\omega_m(g_1 n_1 - g_2 n_2)]^2} \quad (2.1.12)$$

$$n_2 = \frac{\kappa_{e,2} E_R^2}{\kappa_2^2 + [\Delta_2 - 2g_2/\omega_m(g_1 n_1 - g_2 n_2)]^2} \quad (2.1.13)$$

根据标准的输入-输出理论[38] $a_{out}(t) = a_{in}(t) - \sqrt{\kappa_e} a(t)$ 可以得到输出场，其中 $a_{out}(t)$ 是输出场算符。考虑左腔的输出场，有

$$\langle a_{out}(t)\rangle = (E_L - \sqrt{\kappa_{e,1}} a_{S,1})e^{-i\omega_L t} + (E_p - \sqrt{\kappa_{e,1}} a_{1+})e^{-i(\delta+\omega_L)t} - \sqrt{\kappa_{e,1}} a_{1-} e^{i(\delta-\omega_L)t}$$
$$(2.1.14)$$

探测场的透射率可由输出与输入场中在探测频率处的分量振幅之比给出

$$t(\omega_p) = \frac{E_p - \sqrt{\kappa_{e,1}} a_{1+}}{E_p} = 1 - \left[\frac{\kappa_{e,1}}{\kappa_1 + i\Delta_1' - i\delta} - \frac{1}{d(\delta)} \frac{ig_1^2 n_1 \kappa_{e,1}}{(\kappa_1 + i\Delta_1' - i\delta)^2}\right]$$
$$(2.1.15)$$

透射的探测激光束的快速相位色散 $\phi = \arg[t(\omega_p)]$，致使群延迟的 τ_g 表达为

$$\tau_g = \frac{d\varphi}{d\omega_p}\bigg|_{\omega_p=\omega_1} \quad (2.1.16)$$

注意到，如果 $E_R = 0$ 和 $g_2 = 0$，则式(2.1.10)～式(2.1.16)给出熟知的单模腔光力系统的结果，而单模时光力诱导透明和慢光效应已在实验中观察到[16,17]。下面，将从理论上探讨上述考虑的双腔光力系统中发生的这类现象。

2.1.3 结果和讨论

为了说明数值结果,选择真实的双腔光力系统来计算探测场的透射谱。所用参数为[34]:$\omega_1=2\pi\times 205.3$ THz,$\omega_2=2\pi\times 194.1$ THz,$\kappa_1=2\pi\times 520$ MHz,$\kappa_2=1.73$ GHz,$\kappa_{e,1}=0.2\kappa_1$,$\kappa_{e,2}=0.42\kappa_2$,$g_1=2\pi\times 960$ kHz,$g_2=2\pi\times 430$ kHz,$\omega_m=2\pi\times 4$ GHz,$Q_m=87\times 10^3$,其中 Q_m 是纳米机械谐振腔的品质因数,且衰减速率 γ_m 由 ω_m/Q_m 给出。可以看出 $\omega_m>\kappa_1$ 和 $\omega_m>\kappa_2$,因此,该系统处于分离边带区域,也称为好腔极限,这是对机械谐振子基态冷却的必要条件[39]。

光学腔特性可通过两个强泵浦光和一个弱探测光作用来进行研究。先用频率相对于各自腔模红色失谐($\Delta_1=\Delta_2=\omega_m$)的两个泵浦光进行驱动,然后用一个弱探测光对左腔模进行扫描。探测光的透射谱随探测场与腔场之间失谐量 $\Delta_p=\omega_p-\omega_1$ 变化的情况如图 2.1.2 所示,其中 P_L 分别为 0、0.1 μW、1 μW 和 10 μW,而右边泵浦光束功率 P_R 保持为 0.1 μW。当 $P_L=0$ 时,在所探测透射谱的中心出现一个透射凹陷,如图 2.1.2(a)所示。然而,对于 $P_L=0.1$ μW,当探测光与腔频谐振时,宽的腔共振分裂成两个凹陷和一个狭窄的透明窗。随着左泵浦场功率的增加,有效耦合强度 $G_1=g_1\sqrt{n_1}$ 进一步增加,并且与腔谐振时探测场透射谱中透明窗口的宽度也在增加,这一宽度的大小由改进的机械阻尼率 $\gamma_m^{\text{eff}}\approx\gamma_m\left(1+\frac{4g_1^2 n_1}{\kappa_1\gamma_m}+\frac{4g_2^2 n_2}{\kappa_2\gamma_m}\right)$ 给出[16,34,39]。该电磁感应透明现象可由以泵浦光与探测光之间拍频 $\delta=\omega_p-\omega_L$ 振荡的辐射压力解释。如果这种驱动力接近于机械共振频率 ω_m,则振动模相干激发,导致来自强泵浦场光的斯托克斯和反斯托克斯散射。如果在腔的红边带驱动光学腔,则高度非共振斯托克斯散射将被抑制,且仅在腔内有反斯托克斯散射。然而,当探测光与腔发生谐振时,探测场与反斯托克斯场的相消干涉抑制了反斯托克斯场的建立,因而在探测的透射谱中将出现透明窗口。这些过程与线性朗之万方程(2.1.7)~方程(2.1.9)相关。在可分辨边带区域($\kappa_1,\kappa_2<\omega_m$),当泵浦光失谐 $\Delta_1'=\Delta_2'\approx\omega_m$ 时,较低的边带可以忽略,即 $a_{1-}\approx 0$ 和 $a_{2-}\approx 0$[16]。对于左腔内的场的解为

$$a_{1+} = \frac{\sqrt{\kappa_{e,1}} E_p}{-\mathrm{i}x + \kappa_1 + \dfrac{g_1^2 n_1/2}{-\mathrm{i}x + \gamma_m/2 + \dfrac{g_2^2 n_2/2}{-\mathrm{i}x + \kappa_2}}} \tag{2.1.17}$$

式中 $x(=\delta-\omega_m)$ 表示探测频率对于腔频的失谐。当 $g_2=0$ 时,式(2.1.17)回到单一光学腔结果[16],该解具有电磁诱导透明介质对探测场响应的常见形式[19]。参量增强的光机耦合强度 $G_1 = g_1\sqrt{n_1}$ 和 $G_2 = g_2\sqrt{n_2}$ 起到了原子系统中控制光的拉比频率的作用,双光子共振条件由 $\Delta_1' = \Delta_2' = \omega_m$ 给出。

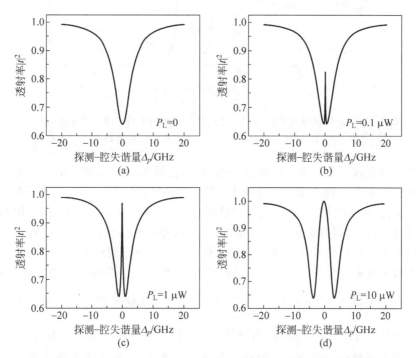

图 2.1.2 以探测场-光学腔失谐量 $\Delta_p = \omega_p - \omega_1$ 为变量的透射谱。左泵浦光功率 P_L 分别为 0、0.1 μW、1 μW 和 10 μW,右泵浦光功率 P_R 保持为 0.1 μW。两个腔被泵浦到它们各自的红边带,即 $\Delta_1 = \omega_m$ 和 $\Delta_2 = \omega_m$

犹如原子中的电磁诱导透明现象,这种效应将引起折射探测光子的显著色散,导致群延迟。对于 $P_L=10$ μW 和 $P_R=0.1$ μW 应用 $\Delta_1=\Delta_2=\omega_m$,图 2.1.3 示出了透射探测光的大小与相位色散随探测-腔失谐 Δ_p 变化的情况。可以清楚地看到,在腔谐振处出现了一个透明窗口,并且伴随急剧的正相位色散,这将导致折射探测光的可调群延迟。另外,透射光的延迟与反射

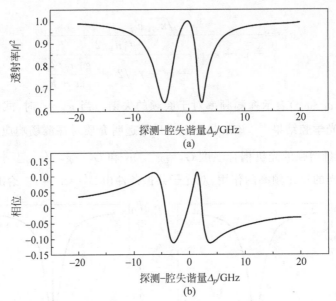

图 2.1.3　透射探测光的(a)大小、(b)相位对于探测-腔失谐 Δ_p 变化的曲线，其中 $P_L=10~\mu\text{W}$ 和 $P_R=0.1~\mu\text{W}$，其他的参数与图 2.1.2 相同

光的超前直接相关。为了验证这一点，我们在图 2.1.4(a)中画出了透射光的群速度延迟 $\tau_g^{(T)}$ 随左泵浦场功率 P_L 变化的情况，其中 $P_R=0.1~\mu\text{W}$ 和 $\Delta_1=\Delta_2=\omega_m$。从图中可以看出，最大传输延迟为 $\tau_g^{(T)}\approx 4.5~\text{ns}$。然而，当关闭右侧的泵浦光时，透射延迟将显著增加，最大延迟竟达 $4~\mu\text{s}$。因此，双模光机系统中的群延迟相比单光学腔系统的延迟会更小。物理起因解释如下：当两个光学腔被泵浦到它们各自的红边带时，机械振子受到左右两个腔的共同影响，总的有效衰减率 $\gamma_m^{\text{eff}}\approx\gamma_m\left(1+\dfrac{4g_1^2 n_1}{\kappa_1\gamma_m}+\dfrac{4g_2^2 n_2}{\kappa_2\gamma_m}\right)$。因此透明窗口的宽度大于在 $g_2=0$ 单腔时的情况，左腔探测透射光中的相位梯度相应变小。因此，对右腔的泵浦将导致透射探测光束的群延迟减少。然而，双腔光力系统中光力诱导透明更加灵活的可控性以及慢光效应在量子存储器[33]及相干光波长转换[34]中潜在的应用，这些在通信系统中是有用的。在图 2.1.4(c)中，我们考虑外部衰减率 $\kappa_{e,1}$ 对群延迟的影响。如果外部衰减在腔衰减中占据主导地位，例如 $\kappa_{e,1}=0.6\kappa$，则最大透射群延迟将进一步增加。此外，图 2.1.4(d)中画出了反射探测光束的群延迟随左泵浦光功率变化的情况，其中延迟是负的，这代表光脉冲的超前。因此，我们可以通过控制泵浦光的功率来调节探测光束的群延迟及超前。

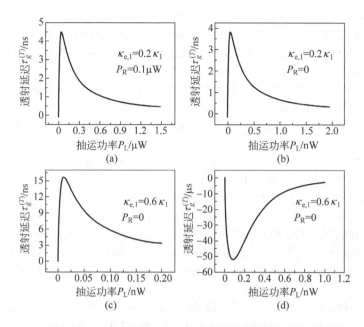

图 2.1.4 (a)~(c)透射；(d)反射探测光束的群延迟 $\tau_g^{(T)}$ 随左泵浦场功率 P_L 变化的关系曲线。考虑了 $\kappa_{e,1}$ 和 p_R 的影响,用 $\Delta_1=\Delta_2=\omega_m$。其他的参数为：$\omega_1=2\pi\times 205.3$ THz,$\omega_2=2\pi\times 194.1$ THz,$\kappa_1=2\pi\times 520$ MHz,$\kappa_2=1.73$ GHz,$\kappa_{e,2}=0.42\kappa_2$,$g_1=2\pi\times 960$ kHz,$g_2=2\pi\times 430$ kHz,$\omega_m=2\pi\times 4$ GHz,$Q_m=87\times 10^3$

2.1.4 小结

本节研究了由两个光学腔耦合于同一个机械振子形成的双腔光力系统在不同驱动条件下的光力诱导透明现象[40]。探测光与反斯托克斯场之间的相消干涉导致在探测透射谱中出现了透明窗口并且伴随着明显正的相位色散,从而产生相应的慢光效应。理论结果显示,透射探测光的延迟时间是可调的,最大可达 4 μs。

参 考 文 献

[1] KIPPENBERG T J, VAHALA K J. Cavity opto-mechanics[J]. Opt. Express, 2007, 15: 17172-17205.

[2] KIPPENBERG T J, VAHALA K J. Cavity optomechanics: back-action at the

mesoscale[J]. Science, 2008, 321: 1172-1176.
[3] MARQUARDT F, GIRVIN S M. Optomechanics[J]. Physics, 2009, 2: 40.
[4] ASPELMEYER M, MEYSTRE P, SCHWA B K. Quantum optomechanics[J]. Phys. Today, 2012, 65: 29-35.
[5] BRENNECKE F, RITTER S, DONNER T, et al. Cavity optomechanics with a Bose-Einstein condensate[J]. Science, 2008, 322: 235-238.
[6] RABL P. Photon blockade effect in optomechanical systems[J]. Phys. Rev. Lett., 2011, 107: 063601.
[7] NUNNENKAMP A, BORKJE K, GIRVIN S M. Single-photon optomechanics[J]. Phys. Rev. Lett., 2011, 107: 063602.
[8] TEUFEL J D, LI D, ALLMAN M S, et al. Circuit cavity electromechanics in the strong-coupling regime[J]. Nature, 2011, 471: 204-208.
[9] VERHAGEN E, DELGLISE S, WEIS S, et al. Quantum-coherent coupling of a mechanical oscillator to an optical cavity mode[J]. Nature, 2012, 482: 63-67.
[10] HE B. Quantum optomechanics beyond linearization[J]. Phys. Rev. A, 2012, 85: 063820.
[11] TEUFEL J D, DONNER T, LI D, et al. Sideband cooling of micromechanical motion to the quantum ground state[J]. Nature, 2011, 475: 359-363.
[12] CHAN J, ALEGRE T P, SAFAVI-NAEINI A H, et al. Laser cooling of a nanomechanical oscillator into its quantum ground state[J]. Nature, 2011, 478: 89-92.
[13] DOBRINDT J M, WILSON-RAE I, KIPPENBERG T J. Parametric normal-mode splitting in cavity optomechanics[J]. Phys. Rev. Lett., 2008, 101: 263602.
[14] GROBLACHER S, HAMMERER K, VANNER M R, et al. Observation of strong coupling between a micromechanical resonator and an optical cavity field[J]. Nature, 2009, 460: 724-727.
[15] AGARWAL G S, HUANG S. Electromagnetically induced transparency in mechanical effects of light[J]. Phys. Rev. A, 2010, 81: 041803.
[16] WEIS S, RIVI`ERE R, DEL'EGLISE S, et al. Optomechanically induced transparency[J]. Science, 2010, 330: 1520-1523.
[17] SAFAVI-NAEINI A H, ALEGRE T P M, CHAN J, et al. Electromagnetically induced transparency and slow light with optomechanics[J]. Nature, 2011, 472: 69-73.
[18] KARUZA M, BIANCOFIORE C, MOLINELLI C, et al. Optomechanically induced transparency in a room temperature membrane-in-the-middle setup[J]. Phys. Rev. A, 2013, 88: 013804.
[19] FLEISCHHAUER M, IMAMOGLU A, MARANGOS J P. Electromagnetically induced transparency: optics in coherent media[J]. Rev. Mod. Phys., 2005, 77: 633-673.
[20] HAU L V, HARRIS S E, DUTTON Z, et al. Light speed reduction to 17 metres

per second in an ultracold atomic gas[J]. Nature, 1999, 397: 594-598.

[21] CHANG D E, SAFAVI-NAEINI A H, HAFEZI M, et al. Slowing and stopping light using an optomechanical crystal array[J]. New J. Phys., 2011, 13: 023003.

[22] BOLLER K J, IMAMOĞLU A, HARRIS S E. Observation of electromagnetically induced transparency[J]. Phys. Rev. Lett., 1991, 66: 2593-2596.

[23] PHILLIPS M C, WANG H, RUMYANTSEV I, et al. Electromagnetically induced transparency in semiconductors via biexciton coherence[J]. Phys. Rev. Lett., 2003, 91: 183602.

[24] LIU N, LANGGUTH L, WEISS T, et al. Plasmonic analogue of electromagnetically induced transparency at the Drude damping limit[J]. Nat. Mater., 2009, 8: 758-762.

[25] SANTORI C, TAMARAT P, NEUMANN P, et al. Coherent population trapping of single spins in diamond under optical excitation[J]. Phys. Rev. Lett., 2006, 97: 247401.

[26] ZHOU X, HOCKE F, SCHLIESSER A, et al. Slowing, advancing and switching of microwave signals using circuit nanoelectromechanics[J]. Nat. Phys., 2013, 9: 179-184.

[27] MASSEL F, HEIKKILÄ T T, PIRKKALAINEN J M, et al. Microwave amplification with nanomechanical resonators[J]. Nature, 2011, 480: 351-354.

[28] HOCKE F, ZHOU X, SCHLIESSER A, et al. Electromechanically induced absorption in a circuit nano-electromechanical system[J]. New J. Phys., 2012, 14: 123037.

[29] DOBRINDT J M, KIPPENBERG T J. Theoretical analysis of mechanical displacement measurement using a multiple cavity mode transducer[J]. Phys. Rev. Lett., 2010, 104: 033901.

[30] LUDWIG M, SAFAVI-NAEINI A H, PAINTER O, et al. Enhanced quantum nonlinearities in a two-mode optomechanical system[J]. Phys. Rev. Lett., 2012, 109: 063601.

[31] KÓMÁR P, BENNETT S D, STANNIGEL K, et al. Single-photon nonlinearities in two-mode optomechanics[J]. Phys. Rev. A, 2013, 87: 013839.

[32] STANNIGEL K, KOMAR P, HABRAKEN S J M, et al. Optomechanical quantum information processing with photons and phonons[J]. Phys. Rev. Lett., 2012, 109: 013603.

[33] QU K, AGARWAL G S. Optical memories and transduction of fields in double cavity optomechanical systems[J]. arXiv:1210.4067 (2012).

[34] HILL J T, SAFAVI-NAEINI A H, CHAN J, et al. Coherent optical wavelength conversion via cavity optomechanics[J]. Nat. Commun., 2012, 3: 1196.

[35] DONG C, FIORE V, KUZYK M C, et al. Optical wavelength conversion via optomechanical coupling in a silica resonator[J]. Ann. Phys., 2015, 527: 100.

[36] GENES C, VITALI D, TOMBESI P, et al. Ground-state cooling of a

micromechanical oscillator: comparing cold damping and cavity-assisted cooling schemes[J]. Phys. Rev. A, 2008, 77: 033804.
[37] BOYD R W. Nonlinear optics[M]. San Diego: Academic Press, 2008.
[38] GARDINER C W, ZOLLER P. Quantum noise[M]. New York: Springer, 2004.
[39] WILSON-RAE I, NOOSHI N, ZWERGER W, et al. Theory of ground state cooling of a mechanical oscillator using dynamical backaction[J]. Phys. Rev. Lett., 2007, 99: 093901.
[40] JIANG C, LIU H X, CUI Y S, et al. Electromagnetically induced transparency and slow light in two-mode optomechanics[J]. Opt. Express, 2013, 21 (10): 12165-12173.

2.2 混杂光力系统中受二能级系统调制的法诺共振和慢光效应

2.2.1 引言

众所周知,介质对弱探测光束的光学响应可通过另外一束强控制场进行调制。电磁诱导透明是一种由不同激发途径之间的量子干涉引起的使不透明物质变得透明的现象[1-3],并产生对称的透明窗口。如果观察电磁诱导透明现象的条件不满足,并且引入了额外的频率失谐量,不对称法诺(Fano)共振就会发生。法诺根据狭窄的离散共振与宽谱线或连续谱的干涉首先解释了法诺共振[4,5]。自从电磁诱导透明和法诺共振被发现以来,它们已在各种量子系统中得到研究,如原子气体、量子阱、量子点、光子晶体、超材料以及超导量子电路[6-13]。电磁诱导透明和法诺共振都已得到各种丰富的应用。电磁诱导透明已被用在慢光[6]、光存储[14]以及非线性过程的增强[15]等方面。此外,法诺共振有许多应用,例如慢光[16]、增强光传输[17]和敏感的生物传感器[18]。值得注意的是,电磁感应透明和欧特莱-汤斯(Autler-Townes)分裂(ATS)都可以在吸收谱或透射谱中出现透明窗口,但它们的物理起源是不同的。电磁诱导透明中的透明是由法诺干涉导致的[4],然而在 Autler-Townes 分裂中是由于电磁泵浦偶极子结构[19]。最近,电磁诱导透明、法诺共振以及 Autler-Townes 分裂都在耦合的回音壁模谐振腔中被观察到[20]。电磁诱导透明和 Autler-Townes 分裂也在超导量子电路中得到理论研究[21]和实验证实[22]。

两个实验[20,22]都观察到了从电磁诱导透明到 Autler-Townes 分裂的过渡。

另外,快速发展的腔光力学领域研究通过辐射压作用下产生的光和机械自由度之间的相互作用[23-25]。在过去十年里,这个领域取得了巨大的成就,包括纳米机械子的基态冷却[26,27]、光力诱导透明(OMIT)[28,29]、慢光[30-32]以及超灵敏传感[33,34]等。此外,法诺共振在不同构型的混合光机械系统中得到研究[35-39]。最近,人们在对含二能级系统[32,36,40-47]或原子系综的混杂光力系统的研究过程中付出了巨大努力[48,49]。

本节将研究文献[45]-文献[47]中讨论的混杂光力系统中的法诺共振和慢光效应,其中二能级系统通过杰尼斯-卡明斯(Jaynes-Cummings,J-C)耦合与机械振子耦合。二能级系统可能是内在的二级缺陷[45]或超导量子比特[50]等。之前的工作表明,耦合腔阵列中的单光子运输可以通过调节腔内二能级系统的跃迁频率[51,52]或是调谐另外一个或两个腔的频率[53]进行控制。反射谱中可以出现某些参数区域的广义法诺线型。本节证实所考虑的混杂光力系统中的吸收波谱可以显示一系列的非对称法诺线型,这一点可以用修饰态的干涉效应来解释。法诺共振可被若干系统参数有效调制,包括跃迁频率、二能级系统的衰减率、量子-机械振子的耦合强度以及腔-控制场的失谐量等。特别地,该混杂系统允许在单法诺共振和双法诺共振间灵活转换。最后,由于透明窗口附近伴随着陡峭的相位色散,我们数值上计算了探测场的群延迟,进而讨论了慢光效应,发现最大群延迟与二能级系统的衰减率成反比。

2.2.2 模型和理论

如图 2.2.1 所示,我们研究的混杂光力系统含有一个单模腔场、一个机械振子和一个二能级系统。机械振子通过辐射压与腔场耦合,它们之间的光力相互作用可用哈密顿量 $H_{om} = -\hbar g a^{\dagger} a (b^{\dagger} + b)$ 描述,$a^{\dagger}(a)$ 和 $b^{\dagger}(b)$ 分别是腔模和机械振子的产生(湮灭)算符;g 是腔场和机械振子间的耦合强度。另外机械振子与二能级系统耦合,可用 J-C 哈密顿量 $H_{J-C} = \hbar J (b^{\dagger} \sigma_{-} + b \sigma_{+})$ 来描述,其中 $\sigma_{-}(\sigma_{+})$ 是二能级系统的升(降)算符,J 表示机械振子和二能级系统间的耦合强度。同时,腔场由频率为 ω_c 的强控制场和频率为 ω_p 的弱探测场驱动。在控制场频率 ω_c 的旋转框架下,这个混杂系统的哈密顿量为

$$H = \hbar\Delta_a a^\dagger a + \hbar\omega_m b^\dagger b + \frac{\hbar}{2}\omega_q \sigma_z + H_{om} + H_{J\text{-}C} + H_{dr} \quad (2.2.1)$$

式中,$\Delta_a = \omega_a - \omega_c$,是腔场频率$\omega_a$与控制场频率$\omega_c$之间的失谐量;$\omega_m$是机械振子的谐振频率;$\omega_q$是二能级系统中基态$|g\rangle$和激发态$|e\rangle$间的跃迁频率,$\sigma_z \equiv |e\rangle\langle e| - |g\rangle\langle g|$是泡利算符。

最后一项H_{dr}代表腔场和驱动场之间的相互作用,可表示为

$$H_{dr} = i\hbar(\Omega a^\dagger - \Omega^* a) + i\hbar(\varepsilon_p e^{-i\Delta t}a^\dagger - \varepsilon_p^* e^{i\Delta t}a) \quad (2.2.2)$$

式中,$\Delta = \omega_p - \omega_c$为探测场和控制场间的失谐量;$\Omega$和$\varepsilon_p$分别为控制场和探测场的拉比频率。

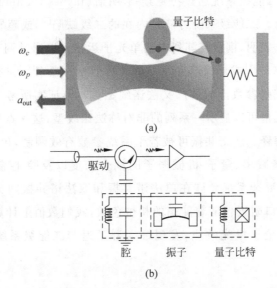

图 2.2.1 混杂光力系统示意图

(a) 光力腔中一个镜子固定,另一个镜子可以振动,相当于一个机械振子,耦合于一个二能级系统(量子比特)。量子比特用振动镜内的红点表示。一个频率为ω_c的强控制场和一个频率为ω_p的弱探测场同时驱动腔场,a_{out}为输出场振幅;(b) 等效的电路图

(请扫Ⅱ页二维码看彩图)

将哈密顿量代入海森伯运动方程并且宏观地引入相应的阻尼项,我们得到

$$\dot{a} = -(\gamma_a + i\Delta_a)a + iga(b^\dagger + b) + \Omega + \varepsilon_p e^{-i\Delta t} \quad (2.2.3)$$

$$\dot{b} = -(\gamma_m + i\omega_m)b + iga^\dagger a + \hbar J\sigma_- \quad (2.2.4)$$

$$\dot{\sigma}_- = -\left(\frac{\gamma_q}{2} + i\omega_q\right)\sigma_- + iJb\sigma_z \quad (2.2.5)$$

$$\dot{\sigma}_z = -\gamma_q(\sigma_z + 1) - 2\mathrm{i}J(b\sigma_+ - b^\dagger\sigma_-) \quad (2.2.6)$$

式中,γ_a、γ_m 和 γ_q 分别为腔场、机械振子和二能级系统的衰减率。因为我们对驱动光学系统对弱探测场的平均响应感兴趣,这里忽略量子和热噪声项。在 $\Omega \gg \varepsilon_p$ 条件下,方程(2.2.3)～方程(2.2.6)可以通过将每个海森伯算符重写为其稳态均值和一个微扰的总和来线性化,即 $a = \alpha + \delta a$,$b = \beta + \delta b$,$\sigma_- = L_0 + \delta\sigma_-$,$\sigma_z = W_0 + \delta\sigma_z$:

$$\dot{\delta a} = -(\gamma_a + \mathrm{i}\Delta_a')\delta a + \mathrm{i}G(\delta b^\dagger + \delta b) + \varepsilon_p \mathrm{e}^{-\mathrm{i}\Delta t} \quad (2.2.7)$$

$$\dot{\delta b} = -(\gamma_m + \mathrm{i}\omega_m)\delta b + \mathrm{i}(G\delta a^\dagger + G^*\delta a) - \mathrm{i}J\delta\sigma_- \quad (2.2.8)$$

$$\dot{\delta\sigma}_- = -\left(\frac{\gamma_q}{2} + \mathrm{i}\omega_q\right)\delta\sigma_- + \mathrm{i}J(\beta\delta\sigma_z + \delta b W_0) \quad (2.2.9)$$

$$\dot{\delta\sigma}_z = -\gamma_q\delta\sigma_z - 2\mathrm{i}J(\beta\delta\sigma_+ + \delta b L_0^* - \beta^*\delta\sigma_- - \delta b^\dagger L_0) \quad (2.2.10)$$

式中,$G = g\alpha$ 为有效光力耦合强度,$\Delta_a' + \Delta_a - g(\beta^* + \beta)$ 为腔场和控制场间的有效失谐量,并且忽略了非线性小量。稳态平均值 α、β、L_0 和 W_0 可由将方程(2.2.3)～方程(2.2.6)所有的时间导数为零获得,满足以下方程:

$$\alpha = \frac{\Omega}{\gamma_a + \mathrm{i}\Delta_a'}, \quad \beta = \frac{\mathrm{i}g|\alpha|^2 - \mathrm{i}JL_0}{\gamma_m + \mathrm{i}\omega_m}$$

$$L_0 = \frac{\mathrm{i}J\beta W_0}{\gamma_q/2 + \mathrm{i}\omega_q}, \quad W_0 = -\frac{\gamma_q^2 + 4\omega_q^2}{\gamma_q^2 + 4\omega_q^2 + 8J^2\beta^2} \quad (2.2.11)$$

这里我们主要考虑的是腔被驱动在红边带的情况(即 $\Delta a \approx \omega_m$)。在 $\omega_m \gg \kappa, G$ 的边带极限中,可以运用旋转波近似(RWA)。通过引入 $\delta a \to \delta a \mathrm{e}^{-\mathrm{i}\Delta t}$,$\delta b \to \delta b \mathrm{e}^{-\mathrm{i}\Delta t}$,$\delta\sigma_- \to \delta\sigma_- \mathrm{e}^{-\mathrm{i}\Delta t}$ 和 $\delta\sigma_z \to \delta\sigma_z \mathrm{e}^{-\mathrm{i}\Delta t}$ 变换到另一个相互作用框架中,得到以下方程:

$$\dot{\delta a} = -(\gamma_a + \mathrm{i}\omega_1)\delta a + \mathrm{i}G\delta b + \varepsilon_p \quad (2.2.12)$$

$$\dot{\delta b} = -(\gamma_m + \mathrm{i}\omega_2)\delta b + \mathrm{i}G^*\delta a - \mathrm{i}J\delta\sigma_- \quad (2.2.13)$$

$$\dot{\delta\sigma}_- = -\left(\frac{\gamma_q}{2} + \mathrm{i}\omega_3\right)\delta\sigma_- + \mathrm{i}J(\beta\delta\sigma_z + \delta b W_0) \quad (2.2.14)$$

$$\dot{\delta\sigma}_z = -(\gamma_q - \mathrm{i}\Delta)\delta\sigma_z - 2\mathrm{i}J(\delta b L_0^* - \beta^*\delta\sigma_-) \quad (2.2.15)$$

式中,$\omega_1 = \Delta_a' - \Delta$,$\omega_2 = \omega_m - \Delta$,$\omega_3 = \omega_q - \Delta$。随后,我们将期望值代入方程(2.2.12)～方程(2.2.15),并且在稳态条件 $\langle\dot{\delta a}\rangle = \langle\dot{\delta b}\rangle = \langle\dot{\delta\sigma}_-\rangle = \langle\dot{\delta\sigma}_z\rangle = 0$

下,有

$$\langle\delta a\rangle = \frac{\Theta\varepsilon_p}{\Gamma_a\Theta + |G|^2[\Gamma_q(\gamma_q - i\Delta) + 2J^2|\beta|^2]} \quad (2.2.16)$$

式中,$\Theta = (\gamma_q - i\Delta)(\Gamma_m\Gamma_q - W_0J^2) + 2iJ^3\beta L_0^* + 2\Gamma_mJ^2|\beta|^2$,$\Gamma_a = \gamma_a + i\omega_1$,$\Gamma_m = \gamma_m + i\omega_2$,$\Gamma_q = \gamma_q/2 + i\omega_3$。根据方程(2.2.11),腔场的稳态光子数$|\alpha|^2$以及机械振子的声子数$|\beta|^2$由以下耦合方程决定:

$$|\alpha|^2\left\{\gamma_a^2 + \left[\Delta_a - \frac{2g^2|\alpha|^2(2\omega_q\varepsilon_1 + \gamma_q\varepsilon_2)}{\varepsilon_1^2 + \varepsilon_2^2}\right]^2\right\} = \Omega^2 \quad (2.2.17)$$

$$|\beta|^2(\varepsilon_1^2 + \varepsilon_2^2)^2 = g^2|\alpha|^4[(2\omega_q\varepsilon_1 + \gamma_q\varepsilon_2)^2 + (\gamma_q\varepsilon_1 - 2\omega_q\varepsilon_2)^2] \quad (2.2.18)$$

式中,$\varepsilon_1 = 2\omega_m\omega_q - \gamma_m\gamma_q + 2J^2W_0$,$\varepsilon_2 = \omega_m\gamma_q + 2\gamma_m\omega_q$,$W_0 = -\frac{\gamma_q^2 + 4\omega_q^2}{\gamma_q^2 + 4\omega_q^2 + 8J^2|\beta|^2}$。

光力腔的输出场可以由输入-输出理论导出[54]

$$\langle a_{\text{out}}\rangle + \Omega + \varepsilon_p e^{-i\Delta t} = 2\gamma_{a,e}\langle a\rangle \quad (2.2.19)$$

式中,$\gamma_{a,e} = \eta\gamma_a$为腔场的外部衰减率。我们感兴趣的是输出场中探测场的频率的分量,并定义$\varepsilon_T = 2\gamma_{a,e}\langle\delta a\rangle/\varepsilon_p = v_p + i\tilde{v}_p$。实部$v_p$和虚部$\tilde{v}_p$分别描述光力系统探测场的吸收和色散行为[28]。另外,探测场频率处的透射系数[55,56]

$$t_p(\omega_p) = \frac{2\gamma_{a,e}\langle\delta a\rangle - \varepsilon_p}{\varepsilon_p} = \frac{2\eta\gamma_a\Theta}{\Gamma_a\Theta + |G|^2[\Gamma_q(\gamma_q - i\Delta) + 2J^2|\beta|^2]} - 1$$

$$(2.2.20)$$

控制场可以改变探头场的传输并引起透明窗口附近快速的相位色散$\phi_t(\omega_p) = \arg[t_p(\omega_p)]$,这会导致明显的群延迟[29,56]

$$\tau_g = \frac{d\phi_t(\omega_p)}{d\omega_p} = \frac{d\{\arg[t_p(\omega_p)]\}}{d\omega_p} \quad (2.2.21)$$

2.2.3 可控的法诺共振

为了证实在此混杂系统中存在法诺共振,我们研究处于探测频率的输出场的吸收谱。在数值分析中,使用了与文献[46]和文献[47]中类似的参数: $\omega_q/(2\pi) = \omega_m/(2\pi) = 100$ MHz,$\gamma_a/(2\pi) = 4$ MHz,$\gamma_m/(2\pi) = 1$ kHz,$\gamma_q/(2\pi) = 0.1$ MHz,$g/(2\pi) = 10$ MHz,$J/(2\pi) = 10$ MHz,$\Omega/(2\pi) = 20$ MHz。

在探测场-腔场失谐量$\Delta_p/(2\pi)$取不同值时,图2.2.2所示是探测场的吸收谱随探测场-腔场失谐量 $\Delta_p/(2\pi)$ 变化的情况。首先考虑两能级系统与机械振子共振的情况,即 $\omega_q = \omega_m$。显然,图 2.2.2(a)左侧的两个线型具有不对称特征,且在 $\Delta_p/(2\pi) = -30$ MHz 和 $\Delta_p/(2\pi) = -10$ MHz 处有两个吸收零点,这就是双法诺共振。这些共振的物理原因与弱探测场和由机械振荡引起的强控制场的光的反斯托克斯散射的相消干涉有关。当腔场-控制场失谐量 $\Delta_a = 1.1\omega_m$ 时,只有一个单一的法诺共振出现在 $\Delta_p/(2\pi) = -20$ MHz。此外,图 2.2.2(a)中中央较宽的吸收峰分为两个对称的吸收峰较窄的透明窗口,可称之为光力诱导透明(OMIT)。当腔场被驱动至红边带,即 $\Delta_a = \omega_m$ 时,可从图 2.2.2(c)看出,有两个几乎对称的透明窗口分别处在 $\Delta_p/(2\pi) = -10$ MHz 和 $\Delta_p/(2\pi) = 10$ MHz。这种效应可以称为双色光力诱导透明[46]。如图 2.2.2(d)和(e)所示,当腔场-控制场的失谐量进一步降低,不对称的单、双法诺共振也可被观察到。

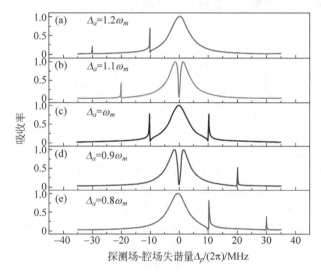

图 2.2.2 探测场-腔场失谐量取不同值时吸收率 v_p 随探测场-腔场失谐量 $\Delta_p = \omega_p - \omega_a$ 变化的曲线

(a) $\Delta_a = 1.2\omega_m$; (b) $\Delta_a = 1.1\omega_m$; (c) $\Delta_a = \omega_m$; (d) $\Delta_a = 0.9\omega_m$; (e) $\Delta_a = 0.8\omega_m$

(请扫 II 页二维码看彩图)

以上提到的单、双法诺共振现象可由基于图 2.2.3 所示的能级图的干涉效应来解释。在我们选择的参数体系中,只考虑单光子和单声子激发。这可以通过方程(2.2.17)和方程(2.2.18)确定的稳态光子数 $|\alpha|^2$ 和稳态声子数 $|\beta|^2$ 进行验证。在没有二能级系统时,光力系统可通过由 $|0_a, 0_m\rangle$、$|0_a, 1_m\rangle$ 和 $|1_a, 0_m\rangle$ 三个态所组成的 Λ 型三能级系统来描述,这里的下标 a 和 m 分别

代表腔和力学模。然而,当一个 $\omega_q=\omega_m$ 的二能级系统与机械振子耦合时,态 $|0_a,1_m\rangle$ 分裂成频率为 ω_m+J 的修饰态 $|0_a,1_m+\rangle$ 和频率为 ω_m-J 的修饰态 $|0_a,1_m-\rangle$[57]。当腔由腔场-控制失谐量为 $\Delta_a=\omega_a-\omega_c$ 的强控制场驱动时,以拍频 Δ 振荡的辐射压力将会产生与控制场相关频率为 $\omega_{as,1}=\omega_c+(\omega_m+J)=\omega_a-\Delta_a+\omega_m+J$ 的斯托克斯和频率为 $\omega_{as,2}=\omega_c+(\omega_m-J)=\omega_a-\Delta_a-J$ 的反斯托克斯散射。反斯托克斯场与腔场接近共振,但斯托克斯场由于与腔场高度非共振而被强烈抑制。频率为 ω_p 的衰减探测场与产生的反斯托克斯之间的干涉引起在探测场吸收光谱中处于 $\Delta_p=\omega_m-\Delta_a+J$ 和 $\Delta_p=\omega_m-\Delta_a-J$ 处的两个透明窗口。我们的定性解释与图 2.2.2 中的数值结果吻合得很好。对于多声子激发情况的分析是相似的,但两修饰态之间的分裂宽度会变宽,因此,法诺共振的位置也会相应改变。

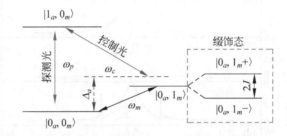

图 2.2.3 二能级系统与机械振子共振时混杂光力系统的能级示意图。一个腔场-控制场失谐量为 $\Delta_a=\omega_a-\omega_c$ 的强控制场被用于腔场,一个弱探测场在用分别用 a 和 m 代表的过渡 $|0_a,0_m\rangle$ 和 $|1_a,0_m\rangle$ 之间扫描。机械振子和二能级系统之间的耦合产生了修饰态 $|0_a,1_m+\rangle$ 和 $|0_a,1_m-\rangle$,这里 $|1_m\pm\rangle=(|1_m,g\rangle\pm|0_m,e\rangle)/\sqrt{2}$

(请扫 II 页二维码看彩图)

我们现在考虑在此混杂系统中二能级系统的过渡频率 ω_q 对法诺共振线形状的影响。图 2.2.4 显示的是腔-控制失谐的蓝失谐(a)ω_q(即 $\omega_q>\omega_m$)和红失谐(b)ω_q(即 $\omega_q<\omega_m$)分别取不同值时吸收光谱对于探测场-腔场的失谐量。在这两种情况下,当腔场-控制场失谐量 Δ_a 从 $1.2\omega_m$ 变化到 $0.8\omega_m$(自上而下)时,我们仍可观察到一系列的非对称法诺线型,但法诺线型的位置与共振情况相比有一个小变化。在蓝失谐情况下($\omega_q-\omega_m>0$),从图 2.2.4(a)可以看到在每个对应的面板中,法诺线型相比图 2.2.2 都发生蓝移。在红失谐情况下,图 2.2.4(b)显示法诺线型的位置分别发生红移。此外,当 $\Delta_a=1.1\omega_m$

时,从图 2.2.4 显然可看到双法诺共振的发生,但图 2.2.2(b)显示只有一个单一的法诺共振和对称的透明窗口。因此,可以通过调整二能级系统的跃迁频率 ω_q 实现在单、双法诺共振之间的转换。特别地,当腔被驱动在其红边带时,对称双色光力诱导透明成为双法诺共振。

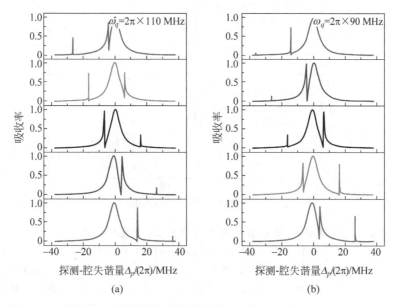

图 2.2.4 吸收率 v_p 随探测场-腔场失谐量 Δ_p 在(a)$\omega_q/(2\pi)=110$ MHz 和 (b)$\omega_q/(2\pi)=90$ MHz 时变化的曲线,其中腔场-控制场失谐量取不同值:$\Delta_a/(2\pi)=1.2\omega_m,1.1\omega_m,\omega_m,0.9\omega_m,0.8\omega_m$(自上而下)

(请扫Ⅱ页二维码看彩图)

此外,光力系统中的法诺共振可被量子比特-机械振子的耦合强度和二能级系统的衰减率调谐。图 2.2.5(a)所示是 $\omega_q=\omega_m$ 和 $\Delta_a=0.9\omega_m$ 而量子比特-机械振子耦合强度取不同值时探测场吸收谱随探测失谐量变化的情况。在这种情况下,透明窗口分别位于 $\Delta_p=\omega_m-\Delta_a+J$ 和 $\Delta_p=\omega_m-\Delta_a-J$。因此,双法诺极小可被量子比特-机械振子的耦合强度调谐。当量子比特-机械振子耦合强度 $J=2\pi\times10$ MHz 时,共振的宽吸收峰分裂成对称透明窗口。因此,此二能级系统中的双法诺共振可通过调节量子比特-机械振子耦合强度 J 成为单法诺共振。在图 2.2.5(b)中,我们讨论了在 $J=2\pi\times8$ MHz 时二能级系统的衰减率对法诺线型的影响。图 2.2.5(b)的插图显示透明窗口的位置保持不变,但随着衰减率 γ_q 的降低,可以观察到更高的最大值和更小的最小值。这种效应与自电离情况下的结果相似,其中最小值取决于辐射效应[4,5]。

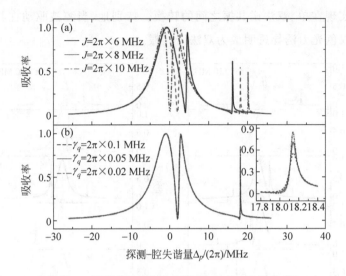

图 2.2.5 吸收率 v_p 随探测场-腔场失谐量 Δ_p 变化的曲线,(a) $\gamma_q = 2\pi \times 0.1$ MHz,量子比特-谐振子耦合强度 J 取不同值,(b) $J = 2\pi \times 8$ MHz,二能级系统的衰减率 γ_q 取不同值。图(b)中的插图是在 $\Delta_p/(2\pi) = 18$ MHz 附近的放大图像。除 $\Delta_a = 0.9\omega_m$ 以外,其他参数与图 2.2.2 相同

(请扫Ⅱ页二维码看彩图)

2.2.4 探测场透射谱中的慢光效应

本节我们首先分析通过这个混杂光力系统的透射探测场的性质。图 2.2.6 所示是探测场的透射率和相位随探测场-腔场失谐量 Δ_p 变化的曲线。当腔场被驱动在其红边带时,从图 2.2.6(a)中可以看出,有两个透明窗口对称地分布在 $\Delta_p/(2\pi) = 10$ MHz 和 $\Delta_p/(2\pi) = 20$ MHz 处。与此同时,如图 2.2.6(b)所示,在透明窗口区域,探测场的相位色散发生急剧的变化。在 $\Delta_p/(2\pi) = \pm 10$ MHz 附近的快速正相色散会导致正的群延迟,也就是所谓的慢光。当腔场-控制场失谐量 $\Delta_a = 0.9\omega_m$ 时,一个对称线形状的透明窗口出现在共振点 $\Delta_p/(2\pi) = 0$ 处,一个法诺线型透明窗口出现在 $\Delta_p/(2\pi) = 20$ MHz。这两种透明窗口都伴随着急剧的相位色散,因此它们可以用来产生

慢光效应。如果我们进一步调节控制域的频率,透明窗口的位置就会改变,因此,在较宽的频率范围内,探测场的群延迟可以有效地调整,且可以忽略损耗。

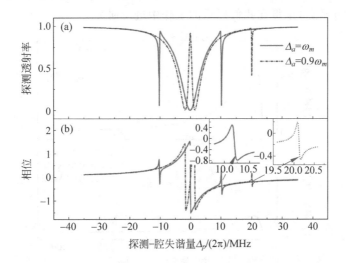

图 2.2.6 探测场的(a)透射率 $|t_p|^2$ 和(b)相位 ϕ_t 分别在 $\Delta_a = \omega_m$ 和 $\Delta_a = 0.9\omega_m$ 时随探测场-腔场失谐量 Δ_p 变化的曲线。图(b)中的插图分别是在 $\Delta_p/(2\pi) = 10$ MHz 和 $\Delta_p/(2\pi) = 20$ MHz 附近的放大图

(请扫Ⅱ页二维码看彩图)

如图 2.2.7 所示,透射探测场的群延迟 τ_g 随控制场的拉比频率在(a)腔场-控制场失谐量 a 和(b)二能级系统的衰减率 γ_q 取不同值时的变化情况。从图 2.2.7(a)可以看出,当拉比频率变化时,群延迟为正值,这代表可以出现慢光效应。如图 2.2.2 所示,这里我们考虑的是在 $\Delta_p/(2\pi) = 10$ MHz,20 MHz,30 MHz 以及 $\Delta_a = \omega_m, 0.9\omega_m, 0.8\omega_m$ 时所对应的慢光效应。当腔场-控制场失谐量 Δ_a 减小时,最大群延迟由 0.55 μs 减小到 0.2 μs,但是最大群延迟的拉比频率增加了。光力系统中的慢光效应的物理原因是,在强控制场的影响下,探测场在通过光力腔时所获得的快速相移。另外,在图 2.2.7(b)中,我们考虑了在 $\Delta_a = \omega_m$ 时二能级系统的衰减率对群延迟的影响。可见最大群延迟与衰减率 γ_q 成反比。如果 $\gamma_q/(2\pi)$ 从 0.1 MHz 减小到 0.05 MHz,最大群延迟 τ_g 就从约 0.55 μs 增加到约 1.1 μs,因此,$\gamma_q/(2\pi) = 0.05$ MHz 时 τ_g 的最大值是 $\gamma_q/(2\pi) = 0.1$ MHz 时的 2 倍。如果衰减率 $\gamma_q/(2\pi)$ 可进一步减小到 0.02 MHz,最大群延迟可被极大增强。最后,我们注意到阿加瓦尔

(Agarwal)等展示了如何应用连接两个低亚稳态的附加耦合场可以将亚光速传播变为超光速光传播[58]。我们也可以应用另一个耦合场控制在混合光机系统光的群延迟,这将在我们以后的工作中进行研究。

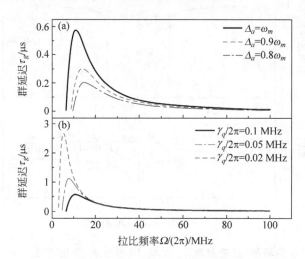

图 2.2.7 群延迟 τ_g 随控制场的拉比频率在(a)腔场-控制场失谐量 a 和(b)二能级系统的衰减率 γ_q 取不同值时的变化曲线

(请扫 Ⅱ 页二维码看彩图)

2.2.5 小结

本节研究了由一个腔场和一个耦合了二能级系统的机械振子组成的混杂光力系统中的法诺共振和慢光效应[59]。研究发现,通过调整腔场和控制腔场之间的失谐量,探测场的吸收谱中可以出现单法诺共振和双法诺共振现象。法诺线型的位置与含有二能级系统的混杂系统的能级图密切相关。另外,本节讨论了跃迁频率和二能级系统的衰减率以及量子比特-机械振子耦合强度对非对称法诺线型的影响。此外,透射探测场中的窄透明窗口和相应的快速相位色散可以引起群延迟,且可以忽略损耗。研究表明,控制场的频率和振幅可以有效地调节探测场的群延迟,并且最大群延迟与二能级系统的衰减率成反比。

参 考 文 献

[1] BOLLER K J, IMAMOGLU A, HARRIS S E. Observation of electromagnetically induced transparency[J]. Phys. Rev. Lett., 1991, 66: 2593.
[2] FLEISCHHAUER M, IMAMOGLU A, MARANGOS J P. Electromagnetically induced transparency: optics in coherent media[J]. Rev. Mod. Phys., 2005, 77:633.
[3] WU Y, YANG X X. Electromagnetically induced transparency in V-, Λ-, and cascade-type schemes beyond steady-state analysis[J]. Phys. Rev. A, 2005, 71: 053806.
[4] FANO U. Effects of configuration interaction on intensities and phase shifts[J]. Phys. Rev., 1961, 124: 1866.
[5] MIROSHNICHENKO A E, FLACH S, KIVSHAR Y S. Fano resonances in nanoscale structures[J]. Rev. Mod. Phys., 2010, 82: 2257.
[6] HAU L V, HARRIS S E, DUTTON Z, et al. Light speed reduction to 17 metres per second in an ultracold atomic gas[J]. Nature, 1999, 397: 594.
[7] PHILLIPS M C, WANG H L, RUMYANTSEV I, et al. Electromagnetically induced yransparency in demiconductors via biexciton voherence[J]. Phys. Rev. Lett., 2003, 91:183602.
[8] CLERK A A, WAINTAL X, BROUWER P W. Fano resonances as a probe of phase coherence in quantum dots[J]. Phys. Rev. Lett., 2001, 86: 4636.
[9] YANG X D, YU M B, KWONG D L, et al. All-optical analog to electromagnetically induced transparency in multiple coupled photonic crystal cavities[J]. Phys. Rev. Lett., 2009, 102: 173902.
[10] GALLI M, PORTALUPI S L, BELOTTI M, et al. Light scattering and Fano resonances in high-Q photonic crystal nanocavities[J]. Appl. Phys. Lett., 2009, 94: 071101.
[11] LIU N, LANGGUTH L, WEISS T, et al. Plasmonic analog of electromagnetically induced transparency at the Drude damping limit[J]. Nat. Mater., 2009, 8: 758.
[12] LUK'YANCHUK B, ZHELUDEV N I, MAIER S A, et al. The Fano resonance in plasmonic nanostructures and metamaterials[J]. Nat. Mater., 2010, 9: 707.
[13] IAN H, LIU Y X, NORI F. Tunable electromagnetically induced transparency and absorption with dressed superconducting qubits[J]. Phys. Rev. A, 2010, 81: 063823.
[14] LIU C, DUTTON Z, BEHROOZI C H, et al. Observation of coherent optical information storage in an atomic medium using halted light pulses[J]. Nature, 2001, 409: 490.

[15] HARRIS S E, FIELD J E, IMAMOGLU A. Nonlinear optical processes using electromagnetically induced transparency[J]. Phys. Rev. Lett., 1990, 64: 1107.

[16] WU C, KHANIKAEV A B, SHVETS G. Broadband slow light metamaterial based on a double-continuum Fano resonance [J]. Phys. Rev. Lett., 2011, 106: 107403.

[17] ZHOU Z K, PENG X N, YANG Z J, et al. Tuning gold nanorod-nanoparticle hybrids into plasmonic Fano resonance for dramatically enhanced light emission and transmission[J]. Nano Lett., 2011, 11: 49.

[18] LEE K L, WU S H, LEE C W, et al. Sensitive biosensors using Fano resonance in single gold nanoslit with periodic grooves[J]. Opt. Express, 2011, 19: 24530.

[19] AUTLER S H, TOWNES C H. Stark effect in rapidly varying fields[J]. Phys. Rev., 1955, 100: 703.

[20] PENG B, ÖZDEMIR Ş K, CHEN W, et al. What is and what is not electromagnetically induced transparency in whispering-gallery microcavities [J]. Nat. Commun., 2014, 5: 5082.

[21] SUN H C, LIU Y X, IAN H, et al. Electromagnetically induced transparency and Autler-Townes splitting in superconducting flux quantum circuits[J]. Phys. Rev. A, 2014, 89: 063822.

[22] LIU Q C, LI T F, LUO X Q, et al. Method for identifying electromagnetically induced transparency in a tunable circuit quantum electrodynamics system [J]. Phys. Rev. A, 2016, 93: 053838.

[23] ASPELMEYER M, KIPPENBERG T J, MARQUARDT F. Cavity optomechanics [J]. Rev. Mod. Phys., 2014, 86: 1391.

[24] MARQUARDT F, GIRVIN S M. Optomechanics[J]. Physics, 2009, 2: 40.

[25] XIONG H, SI L G, LV X Y, et al. Review of cavity optomechanics in the weak-coupling regime: from linearization to intrinsic nonlinear interactions[J]. Sci. China Phys. Mech. Astron., 2015, 58: 1.

[26] CHAN J, ALEGRE T P M, SAFAVI-NAEINI A H, et al. Laser cooling of a nanomechanical oscillator into its quantum ground state [J]. Nature, 2011, 478: 89.

[27] TEUFEL J D, DONNER T, LI D, et al. Sideband cooling of micromechanical motion to the quantum ground state[J]. Nature, 2011, 475: 359.

[28] AGARWAL G S, HUANG S. Electromagnetically induced transparency in mechanical effects of light[J]. Phys. Rev. A, 2010, 81: 041803.

[29] WEIS S, RIVIÈRE R, DELÉGLISE S, et al. Optomechanically induced transparency[J]. Science, 2010, 330: 1520.

[30] SAFAVI-NAEINI A H, ALEGRE T P M, CHAN J, et al. Electromagnetically induced transparency and slow light with optomechanics [J]. Nature, 2011, 472: 69.

[31] ZHOU X, HOCKE F, SCHLIESSER A, et al. Slowing, advancing and switching of microwave signals using circuit nanoelectromechanics[J]. Nat. Phys., 2013, 9: 179.

[32] AKRAM M J, KHAN M M, SAIF F. Tunable fast and slow light in a hybrid optomechanical system[J]. Phys. Rev. A, 2015, 92: 023846.

[33] GAVARTIN E, VERLOT P, KIPPENBERG T J. A hybrid on-chip optomechanical transducer for ultrasensitive force measurements[J]. Nat. Nanotechnol., 2012, 7: 509.

[34] JIANG C, CUI Y, ZHU K D. Ultrasensitive nanomechanical mass sensor using hybrid opto-electromechanical systems[J]. Opt. Express, 2014, 22: 13773.

[35] QU K, AGARWAL G S. Fano resonances and their control in optomechanics[J]. Phys. Rev. A, 2013, 87: 063813.

[36] AKRAM M J, GHAFOOR F, SAIF F. Electromagnetically induced transparency and tunable Fano resonances in hybrid optomechanics[J]. J. Phys. B, 2015, 48: 065502.

[37] YASIRA K A, LIU W M. Controlled electromagnetically induced transparency and Fano resonances in hybrid BEC-optomechanics[J]. Sci. Rep., 2016, 6: 22651.

[38] ZHANG S, LI J, YU R, et al. Optical multistability and Fano line-shape control via mode coupling in whispering-gallery-mode microresonator optomechanics[J]. Sci Rep., 2017, 7: 39781.

[39] AKRAM M J, GHAFOOR F, KHAN M M, et al. Control of Fano resonances and slow light using Bose-Einstein condensates in a nanocavity[J]. Phys. Rev. A, 2017, 95: 023810.

[40] RESTREPO J, FAVERO I, CIUTI C. Fully coupled hybrid cavity optomechanics: quantum interferences and correlations[J]. Phys. Rev. A, 2017, 95: 023832.

[41] COTRUFO M, FIORE A, VERHAGEN E. Coherent atom-phonon interaction through mode field coupling in hybrid optomechanical systems[J]. Phys. Rev. Lett., 2017, 118: 133603.

[42] RESTREPO J, CIUTI C, FAVERO I. Single-polariton optomechanics[J]. Phys. Rev. Lett., 2014, 112: 013601.

[43] PIRKKALAINEN J M, CHO S U, MASSEL F, et al. Cavity optomechanics mediated by a quantum two-level system[J]. Nat. Commun., 2015, 6: 6981.

[44] LECOCQ F, TEUFEL J D, AUMENTADO J, et al. Resolving the vacuum fluctuations of an optomechanical system using an artificial atom[J]. Nat. Phys., 2015, 11: 635.

[45] RAMOS T, SUDHIR V, STANNIGEL K, et al. Nonlinear quantum optomechanics via individual intrinsic two-level defects[J]. Phys. Rev. Lett., 2013, 110: 193602.

[46] WANG H, GU X, LIU Y X, et al. Optomechanical analog of two-color electromagnetically induced transparency: photon transmission through an optomechanical device with a two-level system [J]. Phys. Rev. A, 2014, 90: 023817.

[47] WANG H, GU X, LIU Y X, et al. Tunable photon blockade in a hybrid system consisting of an optomechanical device coupled to a two-level system[J]. Phys. Rev. A, 2015, 92: 033806.

[48] IAN H, GONG Z R, LIU Y X, et al. Cavity optomechanical coupling assisted by an atomic gas[J]. Phys. Rev. A, 2008, 78: 013824.

[49] CHANG Y, SHI T, LIU Y X, et al. Multistability of electromagnetically induced transparency in atom-assisted optomechanical cavities[J]. Phys. Rev. A, 2011, 83: 063826.

[50] LAHAYE M D, SUH J, ECHTERNACH P M, et al. Nanomechanical measurements of a superconducting qubit[J]. Nature, 2009, 459: 960.

[51] ZHOU L, GONG Z R, LIU Y X, et al. Controllable scattering of a single photon inside a one-dimensional resonator waveguide [J]. Phys. Rev. Lett., 2008, 101: 100501.

[52] ZHOU L, DONG H, LIU Y X, et al. Quantum supercavity with atomic mirrors [J]. Phys. Rev. A, 2008, 78: 063827.

[53] LIAO J Q, GONG Z R, ZHOU L, et al. Controlling the transport of single photons by tuning the frequency of either one or two cavities in an array of coupled cavities[J]. Phys. Rev. A, 2010, 81: 042304.

[54] WALLS D F, MILBURN G J. Quantum optics[M]. New York: Springer, 1994.

[55] JIA W Z, WEI L F, LI Y, et al. Phase-dependent optical response properties in an optomechanical system by coherently driving the mechanical resonator[J]. Phys. Rev. A, 2015, 91: 043843.

[56] LIU Y L, WU R, ZHANG J, et al. Controllable optical response by modifying the gain and loss of a mechanical resonator and cavity mode in an optomechanical system [J]. Phys. Rev. A, 2017, 95: 013843.

[57] ORSZAG M. Quantum optics[M]. New York: Springer, 2000.

[58] AGARWAL G S, DEY T N, MENON S. Knob for changing light propagation from subluminal to superluminal[J]. Phys. Rev. A, 2001, 64: 053809.

[59] JIANG C, JIANG L, YU H L, et al. Fano resonance and slow light in hybrid optomechanics mediated by a two-level system [J]. Phys. Rev. A, 2017, 96: 053821.

2.3 机械驱动下宇称-时间-对称的光力系统中的快慢光效应

2.3.1 引言

1998年本德尔(Bender)等证明了具有宇称-时间-对称性($[H,PT]=0$)的非厄米哈密顿量可以有实的本征值[1,2],从此之后宇称-时间-对称的光学受到了广泛的研究[3-7]。此外,调节哈密顿量中的参数可以使宇称-时间-对称的系统经历未破坏的宇称-时间-对称的区域和破坏的宇称-时间-对称的区域之间突然的相变。在奇异点处,一对本征值合并变成复共轭的,可以导致很多奇特的现象,比如损耗诱导的透明[8]、单向不可见性[9,10]、违反左右对称性的功率振荡[11],以及增强的灵敏度[12,13]等。

另外,腔光力学领域还研究电磁和机械自由度之间通过辐射压产生的非线性相互作用[14-16]。基于光力耦合,系统的光学响应可以被机械运动调制,导致光力诱导透明[17-20]、光力诱导吸收[21,22]和光力诱导放大[23]。这些现象的物理原因可以通过辐射压诱导的量子干涉进行解释。此外,光学响应的调制伴随着快速的相位色散,可以进一步用来实现快慢光效应[24-26]。

最近,宇称-时间-对称性和光力系统被提出结合起来构成宇称-时间-对称的光力系统[27-36],其中一个增益腔耦合于一个支持力学模式的损耗腔。这些光力系统中也取得了巨大的进展,包括声子激光[27,28]、宇称-时间-对称性破缺的混沌[29]、探测机械运动时增强的灵敏度[30]、可控的光学响应[31-34]等。此外,刘(Liu)[37]等研究了如何控制由一个损耗腔和一个增益机械振子耦合形成的光力系统中的光学响应,并且他们证明系统中的群速度延迟可以提高几个数量级。

最近的研究表明,光力系统的光学效应可以进一步通过附加的机械驱动进行调制[38-43]。在一束强的光学控制场、一束弱的探测场和一束弱的相干机械驱动场作用下,系统中可以出现更加复杂的量子干涉效应。施加的场可以形成一个封闭的相互作用回路,因此相位依赖的效应非常明显,可以用来控制系统的光学响应[38-43]。值得注意的是,Λ型系统中用来耦合两个较低亚稳态的附加耦合场已经被用来实现快光和慢光效应之间的转变[44]。因此这里的机械驱动场也可以用来控制探测场的群速度延迟。此外,机械驱动场已经在几个光力实验中得到了应用。博赫曼(Bochmann)[45]等在压电光机械晶体中通过电驱动力学模式观察到了电-光力诱导透明,其中力学模式的相位、振幅和频率可以被独立控制。范(Fan)[46]等在多模腔光力学系统中通过同时施加光学泵浦和机械驱动观察到了级联光学透明现象。最近,在光力系统中机械驱动还被用来打破时间-反演-对称性并且实现非互易性的模式转化[47]。

基于以上进展,我们理论上研究了宇称-时间-对称的光力系统在机械驱动时的光学响应。与之前工作不同的是,我们主要讨论了系统在宇称-时间-对称性破缺的区域对弱探测场的响应。我们证实探测透射谱可以通过增益-损耗率、机械驱动的振幅和相位等参数进行有效调制,并且伴随的相位色散的改变导致了透射探测场可控的群速度延迟。

2.3.2 模型和理论

研究的宇称-时间-对称的光力系统如图 2.3.1 所示,其中两个回音壁模式的微型环芯振子通过倏逝场耦合起来,耦合强度为 J。第一个腔是衰减率为 κ_1 的损耗腔,并且通过辐射力耦合于一个共振频率为 ω_m、衰减率为 γ_m 的机械振子。第二个腔是增益率为 κ_2 的增益腔,可以通过在腔中掺杂三价铒离子实现。损耗腔受到频率为 ω_c 的强控制场和频率为 ω_p 的弱探测场驱动。此外,机械振子受到振幅为 ε_m、频率为 $\Omega = \omega_p - \omega_c$、相位为 ϕ_m 的弱的相干声子泵浦场驱动。

在控制场频率 ω_c 的旋转框架下,整个系统的哈密顿量为

$$H = \hbar \Delta_1 a_1^\dagger a_1 + \hbar \Delta_2 a_2^\dagger a_2 + \frac{p^2}{2m} + \frac{1}{2} m \omega_m^2 x^2 - \hbar J (a_1^\dagger a_2 + a_1 a_2^\dagger) -$$

$$\hbar g_1 a_1^\dagger a_1 x + H_{dr} \tag{2.3.1}$$

式中,$a_1 (a_1^\dagger)$ 和 $a_2 (a_2^\dagger)$ 分别是损耗腔和增益腔的湮灭(产生)算符。$\Delta_{1,2} = \omega_{1,2} - \omega_c$ 是共振频率为 $\omega_1 (\omega_2)$ 的两个腔模和控制场之间的失谐量。x 和 p 是有效质量为 m 的机械振子的位移和动量算符。两个腔场之间的耦合强度 J 可以通过它们之间的距离进行控制。第六项表示损耗腔和机械振子之间的相互作用,耦合强度为 g_1。最后一项 H_{dr} 表示驱动场和光力系统之间的相互

第 2 章 腔光力学系统中的慢光效应

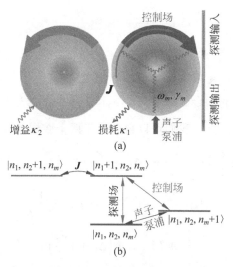

图 2.3.1 (a) 宇称-时间-对称的光力系统示意图；(b) 光力系统的能级图，其中 $n_1(n_2)$ 和 n_m 分别表示损耗腔（增益腔）中的光子数和力学模式中的声子数

（请扫Ⅱ页二维码看彩图）

作用，表示为

$$H_{\mathrm{dr}} = \mathrm{i}\hbar\sqrt{\eta_c\kappa_1}\left[(\varepsilon_c + \varepsilon_p \mathrm{e}^{-\mathrm{i}\Omega t - \mathrm{i}\phi_{pc}})a_1^\dagger - \mathrm{H.c.}\right] - x\varepsilon_m\cos(\Omega t + \phi_m) \tag{2.3.2}$$

式中，ε_c 和 ε_p 分别是控制场和探测场的振幅，它们和功率之间的关系是 $\varepsilon_{c,p} = \sqrt{2P_{c,p}/(\hbar\omega_{c,p})}$。腔 a_1 的总损耗率是 $\kappa_1 = \kappa_{\mathrm{ex},1} + \kappa_{0,1}$，其中 $\kappa_{\mathrm{ex},1}$ 和 $\kappa_{0,1}$ 分别表示外部损耗率和内在损耗率。耦合效率 $\eta_c = \kappa_{\mathrm{ex},1}/\kappa_1$ 可以进行连续调节，在这个工作中我们选择 $\eta_c = 0.4$。$\phi_{pc} = \phi_p - \phi_c$ 是探测场和控制场之间的相对相位。方程(2.3.2)中的最后一项表示机械振子的驱动场作用。

在这个工作中，我们主要研究混杂光力系统对弱探测场的平均响应。因为量子噪声和热噪声的平均值为零，因此可以忽略掉。通过利用海森伯运动方程和诸如 $\langle xa\rangle = \langle x\rangle\langle a\rangle$ 这样的可分解假设，可以得到腔和机械振子算符的平均值方程：

$$\langle\dot{a}_1\rangle = -[\kappa_1/2 + \mathrm{i}(\Delta_1 - g_1\langle x\rangle)]\langle a_1\rangle + \mathrm{i}J\langle a_2\rangle + \sqrt{\eta_c\kappa_1}(\varepsilon_c + \varepsilon_p\mathrm{e}^{-\mathrm{i}\Omega t - \mathrm{i}\phi_{pc}}) \tag{2.3.3}$$

$$\langle\dot{a}_2\rangle = (\kappa_2/2 - \mathrm{i}\Delta_2)\langle a_1\rangle + \mathrm{i}J\langle a_1\rangle \tag{2.3.4}$$

$$\langle \dot{x} \rangle = \frac{\langle p \rangle}{m} \tag{2.3.5}$$

$$\langle \dot{p} \rangle = -m\omega_m^2 \langle x \rangle - \gamma_m \langle p \rangle + \hbar g_1 \langle a_1^\dagger \rangle \langle a_1 \rangle + \varepsilon_m \cos(\Omega t + \phi_m) \tag{2.3.6}$$

其中,唯象地引入了衰减率(κ_1, γ_m)和增益率κ_2。根据方程(2.3.3)~方程(2.3.6),并且在弱耦合区域$g_1 x_{zpf}/\kappa_1 \ll 1$条件下忽略掉光力相互作用,其中$x_{zpf} = \sqrt{\hbar/(2m\omega_m)}$[29],我们可以得到关于两个腔模的非厄米的哈密顿量:

$$\begin{aligned} H_{\text{eff}} &= \left(\Delta_1 - i\frac{\kappa_1}{2}\right)\hbar a_1^\dagger a_1 + \left(\Delta_2 + i\frac{\kappa_2}{2}\right)\hbar a_2^\dagger a_2 - \hbar J(a_1^\dagger a_2 + a_1 a_2^\dagger) \\ &= (a_1^\dagger \ a_2^\dagger) \begin{pmatrix} \hbar\left(\Delta_1 - i\frac{\kappa_1}{2}\right) & -\hbar J \\ -\hbar J & \hbar\left(\Delta_2 + i\frac{\kappa_2}{2}\right) \end{pmatrix} \begin{pmatrix} a_1 \\ a_2 \end{pmatrix} \end{aligned} \tag{2.3.7}$$

如果$\Delta_1 = \Delta_2 = \Delta$,上述哈密顿量很容易对角化,其本征频率为[27,29,30,34,37]

$$\omega_\pm = \Delta - i\frac{\kappa_1 - \kappa_2}{4} \pm \sqrt{J^2 - \left(\frac{\kappa_1 + \kappa_2}{4}\right)^2} \tag{2.3.8}$$

当增益和损耗平衡时($\kappa_1 = \kappa_2$),如果$J > (\kappa_1 + \kappa_2)/4$,则哈密顿量有两个间隔为$\omega_+ - \omega_- = 2\sqrt{J^2 - [(\kappa_1 + \kappa_2)/4]^2}$实的本征频率,系统是宇称-时间-对称的;如果$J < (\kappa_1 + \kappa_2)/4$,则哈密顿量的本征频率是复数,系统工作在宇称-时间对称性破缺的区域。此外,在实际的物理系统中增益率和损耗率不是严格平衡,但是当$J > (\kappa_1 + \kappa_2)/4$时系统可以被认为是宇称-时间对称的[48],且有效衰减率是$(\kappa_1 - \kappa_2)/2$。需要指出的是,在得到方程(2.3.8)时我们忽略了探测场和机械驱动场,因为它们的能量贡献远小于强的控制场,并不会影响系统所给的宇称-时间-对称性。但是当控制场足够强时,光力相互作用会导致相变点的移动。

此外,腔模和力学模式的稳态解可以通过令方程(2.3.3)~方程(2.3.6)的时间求导为零得到,具体如下:

$$a_{1s} = \frac{\sqrt{\eta_c \kappa_1}\, \varepsilon_c (\kappa_2/2 - i\Delta_2)}{(\kappa_1 + i\Delta_1')(\kappa_2/2 - i\Delta_2) - J^2} \tag{2.3.9}$$

$$a_{2s} = -\frac{iJ\sqrt{\eta_c \kappa_1}\, \varepsilon_c}{(\kappa_1 + i\Delta_1')(\kappa_2/2 - i\Delta_2) - J^2} \tag{2.3.10}$$

$$x_s = \frac{\hbar g_1 |a_{1s}|^2}{m\omega_m^2} \tag{2.3.11}$$

$$p_s = 0 \tag{2.3.12}$$

式中,$\Delta_1' = \Delta_1 - g_1 x_s$ 是考虑辐射压效应后腔和控制场之间的有效失谐量。之后,通过令 $\langle a_1 \rangle = a_{1s} + \delta a_1, \langle a_2 \rangle = a_{2s} + \delta a_2, \langle x \rangle = x_s + \delta x, \langle p \rangle = p_s + \delta p$, 方程(2.3.3)~方程(2.3.6)可以被线性化。只保留 δa_1、δa_2、δx、δp 等小量的一阶项,可以得到线性化的海森伯-朗之万方程:

$$\dot{\delta a_1} = -(\kappa_1/2 + i\Delta_1')\delta a_1 + ig_1 a_{1s}\delta x + iJ\delta a_2 + \sqrt{\eta_c \kappa_1}\varepsilon_p e^{-i\Omega t - i\phi_{pc}} \tag{2.3.13}$$

$$\dot{\delta a_2} = (\kappa_2/2 - i\Delta_2)\delta a_2 + iJ\delta a_1 \tag{2.3.14}$$

$$\dot{\delta x} = \frac{\delta p}{m} \tag{2.3.15}$$

$$\dot{\delta p} = -m\omega_m^2 \delta x - \gamma_m \delta p + \hbar g_1(a_{1s}^* \delta a_1 + a_{1s}\delta a_1^+) + \varepsilon_m \cos(\Omega t + \phi_m) \tag{2.3.16}$$

其中被忽略的 $\delta x \delta a_1$ 和 $\delta a_1^+ \delta a_1$ 等项可以引起二阶边带[49]。线性化的朗之万方程(2.3.13)~方程(2.3.16)可以写成矩阵形式

$$\dot{v} = Mv \tag{2.3.17}$$

式中矢量 $v = [\delta \text{Re}[a_1], \delta \text{Im}[a_1], \delta \text{Re}[a_2], \delta \text{Im}[a_2], \delta \tilde{x}, \delta \tilde{p}]^T$,系数矩阵

$$M = \begin{pmatrix} -\kappa_1/2 & \Delta_1' & 0 & -J & -g_0 \text{Im}[a_{1s}] & 0 \\ -\Delta_1' & -\kappa_1/2 & J & 0 & g_0 \text{Re}[a_{1s}] & 0 \\ 0 & -J & \kappa_2/2 & \Delta_2 & 0 & 0 \\ J & 0 & -\Delta_2 & \kappa_2/2 & 0 & 0 \\ 0 & 0 & 0 & 0 & 0 & \omega_m \\ 4g_0 \text{Re}[a_{1s}] & 4g_0 \text{Im}[a_{1s}] & 0 & 0 & -\omega_m & -\gamma_m \end{pmatrix} \tag{2.3.18}$$

式中,$\delta \tilde{x} = \delta x / x_{\text{zpf}}, \delta \tilde{p} = \delta p / p_{\text{zpf}}$ 是无量纲的涨落,其中 $x_{\text{zpf}} = \sqrt{\hbar/(2m\omega_m)}$, $p_{\text{zpf}} = \sqrt{\hbar m \omega_m / 2}$。$\delta \text{Re}[a_i]$ 和 $\delta \text{Im}[a_i]$ 分别是 δa_i 的实部和虚部,$g_0 = g_1 x_{\text{zpf}}$ 是单光子光力耦合强度。因为 ε_p 和 ε_m 比较弱且不会影响系统的稳定性,所以 $\sqrt{\eta_c \kappa_1}\varepsilon_p e^{-i\Omega t - i\phi_{pc}}$ 和 $\varepsilon_m \cos(\Omega t + \phi_m)$ 两项已被忽略。系统只有当方程(2.3.18)中矩阵 M 的所有本征值的实部都小于零时才稳定。稳定性条件解析表达式

可以通过劳斯-赫尔维茨(Routh-Hurwitz)判据[50]得到,但其具体形式过于复杂,这里就不再给出。该工作中所选的参数已数值上验证满足稳定性条件。

为了求解方程(2.3.13)～方程(2.3.16),我们作代换 $\delta o = o_{1+} e^{-i\Omega t} + o_{1-} e^{i\Omega t}$,其中 o 表示 a_1、a_2、x、p 等量。将上述代换代入到方程(2.3.13)～方程(2.3.16),并且比较 $e^{-i\Omega t}$ 和 $e^{i\Omega t}$ 的系数,可以得到

$$a_{1+} = \frac{\sqrt{\eta_c \kappa_1}[1+if(\Omega)]\varepsilon_p e^{-i\phi_{pc}} + ig_1 a_{1s} \chi_m(\Omega)\varepsilon_m e^{-i\phi_m}/2}{\kappa_1/2 + i(\Delta_1' - \Omega) - 2\Delta_1' f(\Omega) - J^2 \chi_c(\omega)} \quad (2.3.19)$$

式中,

$$\chi_m(\Omega) = \frac{1}{m(\omega_m^2 - \Omega^2 - i\gamma_m \Omega)}, \quad \chi_2(\Omega) = \frac{1}{\kappa_2/2 + i\Delta_2 + i\Omega}$$

$$f(\Omega) = \frac{\hbar g_1^2 |a_{1s}|^2 \chi_m(\Omega)}{\kappa_1/2 - i\Delta_1' - i\Omega - J^2 \chi_2(\Omega)}$$

$$\chi_c(\Omega) = \chi_2^*(-\Omega) + if(\Omega)[\chi_2^*(-\Omega) - \chi_2(\Omega)] \quad (2.3.20)$$

根据输入-输出关系[51] $\langle a_{1,\text{out}}(t)\rangle = \langle a_{1,\text{in}}(t)\rangle - \sqrt{\eta_c \kappa_1} a_1(t)$,得到在原始框架下光力腔的输出场为

$$\langle a_{1,\text{out}}(t)\rangle = (\varepsilon_c - \sqrt{\eta_c \kappa_1} a_{1s}) e^{-i\omega_c t} + (\varepsilon_p e^{-i\phi_{pc}} - \sqrt{\eta_c \kappa_1} a_{1+}) e^{-i(\Omega+\omega_c)t} - \sqrt{\eta_c \kappa_1} a_{1-} e^{-(\omega_c - \Omega)t}$$

$$= (\varepsilon_c - \sqrt{\eta_c \kappa_1} a_{1s}) e^{-i\omega_c t} + (\varepsilon_p e^{-i\phi_{pc}} - \sqrt{\eta_c \kappa_1} a_{1+}) e^{-i\omega_p t} - \sqrt{\eta_c \kappa_1} a_{1-} e^{-(2\omega_c - \omega_p)t} \quad (2.3.21)$$

从方程(2.3.21)可以看出,输出场中包含两个输入分量,即频率为 ω_c 的控制场和频率为 ω_p 的探测场,还包括新产生的频率为 $2\omega_c - \omega_p$ 的四波混频分量。探测的透射率定义为

$$t_p = (\varepsilon_p e^{-i\phi_{pc}} - \sqrt{\eta_c \kappa_1} a_{1+})/(\varepsilon_p e^{-i\phi_{pc}}) = t_1 + t_2 \quad (2.3.22)$$

式中,

$$t_1 = 1 - \frac{1+if(\Omega)}{\kappa_1/2 + i(\Delta_1' - \Omega) - 2\Delta_1' f(\Omega) - J^2 \chi_c(\omega)} \eta_c \kappa_1 \quad (2.3.23)$$

$$t_2 = -\frac{ig_1 a_{1s} \chi_m(\Omega)\varepsilon_m/2\varepsilon_p}{\kappa_1/2 + i(\Delta_1' - \Omega) - 2\Delta_1' f(\Omega) - J^2 \chi_c(\omega)} \sqrt{\eta_c \kappa_1} e^{-i\phi}$$

$$(2.3.24)$$

其中，$\phi=\phi_m-\phi_{pc}$ 是机械驱动场和施加的光场之间的相位差。t_1 是没有机械泵驱动时探测场透射谱的表达式，当腔之间耦合强度 $J=0$ 时 t_1 可以化简到典型的光力系统中的结果[18]。t_2 代表的是机械驱动的贡献，会对探测场透射谱带来改变。探测场透射谱主要由 t_1 和 t_2 之间的干涉效应决定，其中相位差 ϕ 是一个重要的调节参数。此外，探测透射谱的调制会伴随着相位色散 $\phi_t(\omega_p)=\arg[t_p(\omega_p)]$ 的改变，进而导致可调的群速度延迟

$$\tau_g=\frac{\mathrm{d}\phi_t(\omega_p)}{\mathrm{d}\omega_p}=\frac{\mathrm{d}\{\arg[t_p(\omega_p)]\}}{\mathrm{d}\omega_p} \tag{2.3.25}$$

2.3.3 机械驱动调制的探测场透射谱

接下来，我们根据实验上可行的参数数值模拟不同条件下探测场的透射谱[10,52]，具体参数包括：损耗腔的频率 $\omega_1=2\pi c/\lambda$，$\lambda=1550$ nm，半径 $R=15$ μm，$g_1=\omega_1/R$，$\kappa_1/(2\pi)=6$ MHz，$m=6.2$ ng，$\omega_m/(2\pi)=78$ MHz，$\gamma_m/(2\pi)=12$ kHz，$\eta_c=0.4$，$J=0.2(\kappa_1+\kappa_2)$。因为腔 a_2 是一个增益腔且耦合于损耗腔 a_1，系统在某些参数区域可能不稳定。为了研究可调的快慢光效应，系统应工作于稳定区域。图 2.3.2 所示是随 κ_2/κ_1 和 $J/(\kappa_1+\kappa_2)$ 变化时的稳定图，从图中可以看出稳定区域随着腔之间耦合强度 J 的增强而变大。当 $J/(\kappa_1+\kappa_2)$ 大于转变点 0.25 时，只要 $\kappa_2<\kappa_1$，则系统处于宇称-时间-对称的区域。

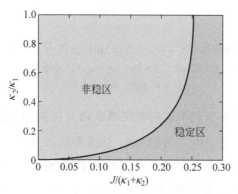

图 2.3.2 关于 κ_2/κ_1 和 $J/(\kappa_1+\kappa_2)$ 的稳定图

(请扫 Ⅱ 页二维码看彩图)

我们首先研究在没有机械驱动时增益-损耗率对探测场透射谱的影响,然后再讨论机械驱动的影响。图 2.3.3 所示是增益-损耗率 κ_2/κ_1 取不同值时 (a)探测场透射率 $|t_p|^2$ 和(b)相位色散 ϕ_t 随失谐量 $(\Omega-\omega_m)/\kappa_1$ 变化的情况。当 $\kappa_2/\kappa_1=-0.1$,即腔 a_2 是损耗腔时,从图 2.3.3(a)可以看出,在 $\Omega=\omega_m$ 处 $|t_p|^2\approx 0.4$。但是当 $\kappa_2/\kappa_1=10^{-4}$,即腔 a_2 变成一个增益腔时,在 $\Omega=\omega_m$ 处 $|t_p|^2\approx 1$。因此,腔 a_2 的增益和辐射压诱导的干涉效应导致在较低控制场功率时出现了光力诱导透明现象。进一步增加增益-损耗率 κ_2/κ_1 到 0.1,探测透射率 $|t_p|^2$ 可以大于 1,代表透射探测场被放大。此外,图 2.3.3(a)中增益-损耗率引起探测透射率 $|t_p|^2$ 的变化伴随着相位色散的改变。从图 2.3.3(b)可以看出,当腔 a_2 是增益腔时,相位在 $\Omega=\omega_m$ 处的斜率比两个腔都是损耗腔时的斜率大,且随着增益-损耗率增加而变大。因此,透射探测场的群速度延迟可以通过增益-损耗率调制,将在后面进行讨论。

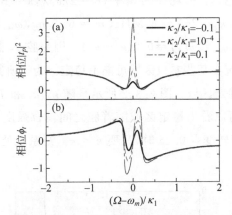

图 2.3.3 (a)探测场透射率 $|t_p|^2$ 和(b)相位色散 ϕ_t 随失谐量 $(\Omega-\omega_m)/\kappa_1$ 变化的曲线。除了 $J=0.2(\kappa_1+\kappa_2)$ 外其他参数与图 2.3.2 相同

(请扫Ⅱ页二维码看彩图)

接下来讨论增益-损耗率固定时机械驱动对探测场透射率 $|t_p|^2$ 的影响。图 2.3.4 所示是 $\kappa_2/\kappa_1=0.1$ 时 $\Omega=\omega_m$ 处探测场透射率 $|t_p|^2$ 随相位差 ϕ 和机械驱动的振幅 ε_m 变化的情况。可以看出,随着振幅 ε_m 的增大,相位依赖的效应非常明显。当 ε_m 固定时,探测透射率 $|t_p|^2$ 在 $\phi=\pi/2$ 附近有最大值,而在 $\phi=3\pi/2$ 附近有最小值。此外,当 $\phi=\pi/2$ 时,透射率 $|t_p|^2$ 随着振幅 ε_m 的增加单调变大,但是当 $\phi=3\pi/2$ 时,$|t_p|^2$ 在 $\varepsilon_m=6$ pN 附近有一个极小值。

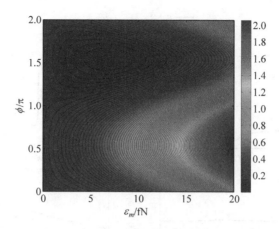

图 2.3.4　$\kappa_2/\kappa_1=0.1$ 时 $\Omega=\omega_m$ 处探测场透射率 $|t_p|^2$ 随相位差 ϕ 和机械驱动的振幅 ε_m 变化的等高线图。因为探测场比控制场弱很多，该工作中取 $\varepsilon_p=\varepsilon_c/1000$。其他参数与图 2.3.3 相同

（请扫Ⅱ页二维码看彩图）

为了能更加清晰地看出机械驱动的影响，我们在图 2.3.5 中画出了不同机械驱动下探测场透射率 $|t_p|^2$ 和相位 ϕ_t 随探测-控制失谐量变化的情况。从图 2.3.3(a) 和 (c) 可以看出，没有机械驱动时 ($\varepsilon_m=0$)，在 $\Omega=\omega_m$ 处 $|t_p|^2\approx 3.5$。当机械驱动施加到机械振子上且振幅 $\varepsilon_m=2$ pN 时，如果相位差 $\phi=\pi/2$，则探测场透射率增加到 $|t_p|^2\approx 6.2$；如果相位差 $\phi=3\pi/2$，则探测场透射率抑制到 $|t_p|^2\approx 1.5$。此外，图 2.3.5(d) 表明，$\varepsilon_m=2$ pN，$\phi=\pi/2$ 时相位色散 ϕ_t 的斜率为正值，且大于 $\varepsilon_m=0$ 时对应的斜率，这可以引起可调的慢光效应。如果相位差 $\phi=3\pi/2$，相位色散 ϕ_t 在 $\Omega=\omega_m$ 处的斜率变为负值，这表示可以出现快光效应。因此，调节机械驱动的相位可以实现正常色散和反常色散之间的转变，从而实现可调的快慢光效应。

探测场透射谱中相位依赖的关系可以由方程 (2.3.22) ～方程 (2.3.23) 中 t_1 和 t_2 之间的干涉效应进行解释。图 2.3.6 所示是相位差分别为 (a) $\phi=\pi/2$ 和 (b) $\phi=3\pi/2$ 时 $|t_1|^2$、$|t_2|^2$、$|t_p|^2$ 随振幅 ε_m 变化的情况。红色虚线代表的是没有机械驱动时的探测场透射率 $|t_1|^2$，因此当 ε_m 变化时 $|t_1|^2$ 保持不变。$|t_2|^2$ 代表的是机械驱动的贡献，并且随着 ε_m 的增强而单调增加。$|t_1|^2$ 和 $|t_2|^2$ 之间的干涉可以通过相位差 ϕ 进行灵活的控制。当 $\phi=\pi/2$ 时，$|t_1|^2$

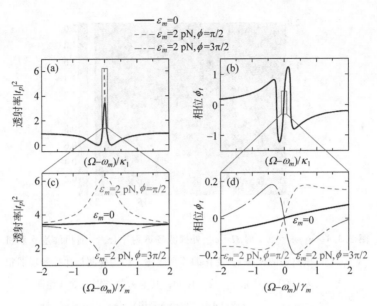

图 2.3.5　不同机械驱动下探测场透射率 $|t_p|^2$((a)和(c))和相位 ϕ_t((b)和(d))随探测-控制失谐量变化的曲线

(请扫Ⅱ页二维码看彩图)

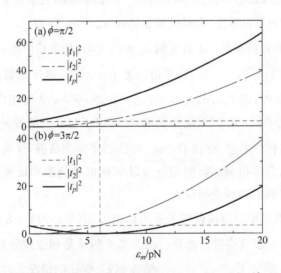

图 2.3.6　相位差分别为(a) $\phi=\pi/2$ 和(b) $\phi=3\pi/2$ 时 $|t_1|^2$、$|t_2|^2$、$|t_p|^2$ 随振幅 ε_m 变化的曲线

(请扫Ⅱ页二维码看彩图)

和$|t_2|^2$之间发生相长干涉,因此$|t_p|^2$比$|t_1|^2$和$|t_2|^2$都要大,并且随ε_m单调增大。在$\varepsilon_m=6$ pN附近$|t_1|^2$和$|t_2|^2$之间发生完全相长干涉,这时$|t_p|^2\approx 4|t_1|^2=4|t_2|^2$。但是,如果相位差$\phi=3\pi/2$,$|t_1|^2$和$|t_2|^2$之间发生相消干涉。探测场透射率$|t_p|^2$从$\varepsilon_m=0$时的3.5减小到$\varepsilon_m=6$ pN时的0,这时发生了完全相消干涉。进一步增加ε_m,$|t_p|^2$又开始增加并且超过了初始值。因此,探测场透射率$|t_p|^2$可以通过机械驱动进行调制并且依赖于相位差ϕ。

2.3.4 透射探测场中可控的慢光和快光效应

基于上述讨论,我们发现探测场透射率$|t_p|^2$和相位色散ϕ_t可以通过增益-损耗率和机械驱动进行控制,从而引起可调的群速度延迟τ_g。本节我们主要研究在不同条件下怎样控制群速度延迟τ_g。图2.3.7所示是(a)增益-损耗率和(b)机械驱动取不同值时群速度延迟τ_g随控制场功率变化的情况。当$\kappa_2/\kappa_1=-0.1$时,即腔a_2也是损耗腔,从图2.3.7(a)可以看出,群速度延迟τ_g为正值,且随着控制场功率P_c增加时$\tau_g<3$ μs。但当腔a_2变成增益腔时,比如$\kappa_2/\kappa_1=0.1$,群速度延迟τ_g为正值且$P_c<6$ μW时τ_g最大值约为0.6 ms。如果P_c进一步增加,τ_g变为负值。因此,在宇称-时间-对称的光力系统中的最大群速度延迟提高了大约两个数量级,并且通过调节控制场的功率可以实现可调的慢光和快光。如果增益-损耗率增大,最大群速度延迟可以被进一步延长。此外,图2.3.7(b)表明,快慢光效应还可以由机械驱动调控。当控制场功率$P_c<4$ μW时,我们可以通过调节相位差ϕ等于$\pi/2$或者$3\pi/2$实现快慢光效应之间的转变。光学控制场和机械驱动场的影响可以从图2.3.7(b)看得更加清晰。在没有机械驱动时,群速度延迟τ_g为正,且在$P_c<4$ μW时随着控制场功率单调增加。当机械驱动场也加上时,更加复杂的量子干涉效应出现,导致群速度延迟τ_g发生变化。如果$\phi=\pi/2$,τ_g比没有机械驱动时的值要大;如果$\phi=3\pi/2$,τ_g变为负值且可由控制场功率P_c调制。

图2.3.8所示是群速度延迟τ_g随机械驱动场的振幅和相位变化的情况。从图2.3.8(a)可以看出,当$\phi=\pi/2$时,群速度延迟τ_g总是为正且随着振幅ε_m增加而单调变长。但是当$\phi=3\pi/2$时,在$\varepsilon_m=6$ pN附近可以实现快光

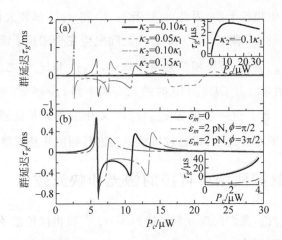

图 2.3.7　(a) 增益-损耗率 κ_2/κ_1，(b) 机械驱动的振幅 ε_m 和相位差 ϕ 取不同值时群速度延迟 τ_g 随控制场功率 P_c 变化的曲线

(请扫Ⅱ页二维码看彩图)

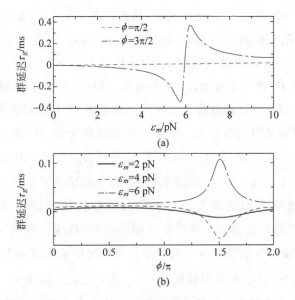

图 2.3.8　群速度延迟 τ_g 随 (a) 机械驱动的振幅 ε_m 和 (b) 相位差 ϕ 变化的曲线

(请扫Ⅱ页二维码看彩图)

($\tau_g<0$)和慢光($\tau_g>0$)之间的转变。此外，图 2.3.8(b) 展示的是不同机械驱动振幅 ε_m 时群速度延迟 τ_g 对相位差 ϕ 的依赖关系。从图中可以看出，当 $\varepsilon_m=4$ pN 和 $\varepsilon_m=6$ pN 时改变相位差可以实现慢光到快光的转变，并且群速

度延迟 τ_g 的最小值出现在 $\phi=3\pi/2$ 附近。如果机械驱动的振幅 ε_m 增加到 8 pN，群速度延迟 τ_g 总为正值，从而只有慢光效应。因此，该光力系统中通过调节机械驱动的振幅和相位可以实现慢光和快光效应之间的转变。

最后，我们研究两个腔之间的耦合强度对探测场透射率 $|t_p|^2$ 和群速度延迟 τ_g 的影响。图 2.3.9(a) 和 (b) 所示是 $\kappa_2/\kappa_1=0.1$ 时 $|t_p|^2$ 和 τ_g 随 $J/(\kappa_1+\kappa_2)$ 变化的情况。从图 2.3.9(a) 可以看出透射率 $|t_p|^2$ 在 $J/(\kappa_1+\kappa_2)=0.145$ 附近有一个峰值，从而导致最大群速度延迟约为 0.45 ms。从图 2.3.2 可以看出，$J/(\kappa_1+\kappa_2)=0.145$ 是 $\kappa_2/\kappa_1=0.1$ 时稳定区和非稳区的转变点。随着进一步增加耦合强度 J，透射率 $|t_p|^2$ 和群速度延迟 τ_g 开始变小。对于这里考虑的增益-损耗非平衡的情况（$\kappa_2/\kappa_1=0.1$），系统在宇称-时间-对称性破缺的区域仍可以稳定。在宇称-时间-对称的区域（$J/(\kappa_1+\kappa_2)>0.25$），通过调节机械驱动的相位仍可以实现慢光和快光之间的转变。对于增益和损耗平衡的情况（图 2.3.9(c) 和 (d)），最大透射率和群速度延迟 τ_g 存在于相变点 $J/(\kappa_1+\kappa_2)=0.25$ 附近。最近金扎克（Goldzak）等在宇称时间-对称的光学波导中发现奇异点处的群速度可以达到零[53]，而图 2.3.9(d) 表明群速度延迟在宇称-时间-对称的光力系统中奇异点附近可以被大大增强。

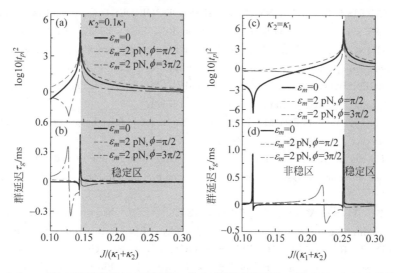

图 2.3.9　ε_m 和 ϕ 取不同值时(a)透射率 $|t_p|^2$ 和(b)群速度延迟 τ_g 随 $J/(\kappa_1+\kappa_2)$ 变化的曲线

（请扫Ⅱ页二维码看彩图）

2.3.5 小结

本节研究了宇称-时间-对称的光力系统在一束强的控制场、一束弱的探测场和一个机械驱动场同时作用下的光学响应[54]。光学增益导致探测场透射率在控制场功率较低时可以大于1，并且可以进一步通过机械驱动场的振幅和相位进行控制。同时，透射探测场的相位色散也发生了改变，导致可调的慢光和快光效应。我们证实透射探测场的群速度延迟可以通过几个参数进行调制，包括增益-损耗率、控制场功率、机械驱动的振幅和相位等。尤其重要的是，宇称-时间-对称的光力系统中的群速度延迟相比于不包含增益的光力系统中的群速度延迟要高两个数量级。

参 考 文 献

[1] BENDER C M, BOETTCHER S. Real spectra in non-Hermitian Hamiltonians having PT symmetry[J]. Phys. Rev. Lett., 1998, 80: 5243.

[2] BENDER C M, BRODY D C, JONES H F. Complex extension of quantum mechanics[J]. Phys. Rev. Lett., 2002, 89: 270401.

[3] MAKRIS K G, GANAINY R E, CHRISTODOULIDES D N, et al. Beam dynamics in PT symmetric optical lattices[J]. Phys. Rev. Lett., 2008, 100: 103904.

[4] CHONG Y D, GE L, STONE A D. PT-symmetry breaking and laser-absorber modes in optical scattering systems[J]. Phys. Rev. Lett., 2011, 106: 093902.

[5] REGENSBURGER A, BERSCH C, MIRI M A, et al. Parity-time synthetic photonic lattices[J]. Nature, 2012, 488: 167171.

[6] FENG L, WONG Z J, MA R M, et al. Single-mode laser by parity-time symmetry breaking[J]. Science, 2014, 346: 972-975.

[7] CHANG L, JIANG X, HUA S, et al. Parity-time symmetry and variable optical isolation in active-passive-coupled microresonators[J]. Nat. Photon., 2014, 8: 524-529.

[8] GUO A, SALAMO G J, DUCHESNE D, et al. Observation of PT-symmetry breaking in complex optical potentials[J]. Phys. Rev. Lett., 2009, 103: 093902.

[9] LIN Z, RAMEZANI H, EICHELKRAUT T, et al. Unidirectional invisibility induced by PT-symmetric periodic structures [J]. Phys. Rev. Lett., 2011, 106: 213901.

[10] PENG B, ÖZDEMIR Ş K, LEI F, et al. Parity-time-symmetric whispering-gallery microcavities[J]. Nat. Phys., 2014, 10: 394-398.

[11] RÖTER C E, MAKRIS K G, GANAINY R E, et al. Observation of parity-time symmetry in optics[J]. Nat. Phys., 2010, 6: 192-195.

[12] CHEN W, ÖZDEMIR Ş K, ZHAO G, et al. Exceptional points enhance sensing in an optical microcavity[J]. Nature, 2017, 548: 192-196.

[13] HODAEI H, HASSAN A U, WITTEK S, et al. Enhanced sensitivity at higher-order exceptional points[J]. Nature, 2017, 548: 187-191.

[14] ASPELMEYER M, KIPPENBERG T J, MARQUARDT F. Cavity optomechanics [J]. Rev. Mod. Phys., 2014, 86: 1391-1452.

[15] MARQUARDT F, GIRVIN S M. Optomechanics[J]. Physics, 2009, 2: 40.

[16] XIONG H, SI L G, LÜ X Y, et al. Review of cavity optomechanics in the weak-coupling regime: from linearization to intrinsic nonlinear interactions[J]. Sci. China Phys. Mech. Astron., 2015, 58: 1-13.

[17] AGARWAL G S, HUANG S. Electromagnetically induced transparency in mechanical effects of light[J]. Phys. Rev. A, 2010, 81: 041803.

[18] WEIS S, RIVIÈRE R, DELÉGLISE S, et al. Optomechanically induced transparency[J]. Science, 2010, 330: 1520-1523.

[19] SAFAVI-NAEINI A H, ALEGRE T P M, CHAN J, et al. Electromagnetically induced transparency and slow light with optomechanics[J]. Nature, 2011, 472: 69-73.

[20] JIANG C, LIU H, CUI Y, et al. Electromagnetically induced transparency and slow light in two-mode optomechanics[J]. Opt. Express, 2013, 21: 12165-12173.

[21] SINGH V, BOSMAN S J, SCHNEIDER B H, et al. Optomechanical coupling between a multilayer graphene mechanical resonator and a superconducting microwave cavity[J]. Nat. Nanotechnol., 2014, 9: 820-824.

[22] JIANG C, CUI Y, BIAN X, et al. Phase-dependent multiple optomechanically induced absorption in multimode optomechanical systems with mechanical driving [J]. Phys. Rev. A, 2016, 94: 023837.

[23] MASSEL F, HEIKKILÄ T T, PIRKKALAINEN J M, et al. Microwave amplification with nanomechanical resonators[J]. Nature, 2011, 480: 351-354.

[24] CHEN B, JIANG C, ZHU K D. Slow light in a cavity optomechanical system with a Bose-Einstein condensate[J]. Phys. Rev. A, 2011, 83: 055803.

[25] AKRAM M J, KHAN M M, SAIF F. Tunable fast and slow light in a hybrid optomechanical system[J]. Phys. Rev. A, 2015, 92: 023846.

[26] ZHOU X, HOCKE F, SCHLIESSER A, et al. Slowing, advancing and switching of microwave signals using circuit nanoelectromechanics[J]. Nat. Phys., 2013, 9: 179-184.

[27] JING H, ÖZDEMIR S K, LÜ X Y, et al. PT-symmetric phonon laser[J]. Phys. Rev. Lett., 2014, 113: 053604.

[28] HE B, YANG L, XIAO M. Dynamical phonon laser in coupled active-passive

microresonators[J]. Phys. Rev. A, 2016, 94: 031802.

[29] LÜ X Y, JING H, MA J Y, et al. PT-symmetry-breaking chaos in optomechanics [J]. Phys. Rev. Lett., 2015, 114: 253601.

[30] LIU Z P, ZHANG J, ÖZDEMIR S K, et al. Metrology with PT-symmetric cavities: enhanced sensitivity near the PT-phase transition[J]. Phys. Rev. Lett., 2016, 117: 110802.

[31] JING H, ÖZDEMIR S K, GENG Z, et al. Optomechanically-induced transparency in parity-time-symmetric microresonators[J]. Sci. Rep., 2015, 5: 9663.

[32] LI W, JIANG Y, LI C, et al. Parity-time-symmetry enhanced optomechanically-induced-transparency[J]. Sci. Rep., 2016,6: 31095.

[33] JIAO Y, LÜ H, QIAN J, et al. Nonlinear optomechanics with gain and loss: amplifying higher-order sideband and group delay[J]. New J. Phys., 2016, 18: 083034.

[34] ZHANG X Y, GUO Y Q, PEI P, et al. Optomechanically induced absorption in parity-time-symmetric optomechanical systems [J]. Phys. Rev. A, 2017, 95: 063825.

[35] SCHÖNLEBER D W, EISFELD A, GANAINY R E. Optomechanical interactions in non-Hermitian photonic molecules[J]. New J. Phys., 2016, 18: 045014.

[36] JING H, ÖZDEMIR S K, LÜ H, et al. High-order exceptional points in optomechanics[J]. Sci. Rep., 2017, 7: 3386.

[37] LIU Y L, WU R, ZHANG J, et al. Controllable optical response by modifying the gain and loss of a mechanical resonator and cavity mode in an optomechanical system [J]. Phys. Rev. A, 2017, 95: 013843.

[38] JIA W Z, WEI L F, LI Y, et al. Phase-dependent optical response properties in an optomechanical system by coherently driving the mechanical resonator[J]. Phys. Rev. A, 2015, 91: 043843.

[39] MA J Y, YOU C, SI L G, et al. Optomechanically induced transparency in the presence of an external time-harmonic-driving force[J]. Sci. Rep., 2015,5: 11278.

[40] XU X W, LI Y. Controllable optical output fields from an optomechanical system with mechanical driving[J]. Phys. Rev. A, 2015, 92: 023855.

[41] SUZUKI H, BROWN E, STERLING R. Nonlinear dynamics of an optomechanical system with a coherent mechanical pump: second-order sideband generation[J]. Phys. Rev. A, 2015, 92: 033823.

[42] LI Y, HUANG Y Y, ZHANG X Z, et al. Optical directional amplification in a three-mode optomechanical system[J]. Opt. Express, 2017, 25: 18907-18916.

[43] SI L G, XIONG H, ZUBAIRY M S, et al. Optomechanically induced opacity and amplification in a quadratically coupled optomechanical system[J]. Phys. Rev. A, 2017, 95: 033803.

[44] AGARWAL G S, DEY T N, MENON S. Knob for changing light propagation

from subluminal to superluminal[J]. Phys. Rev. A, 2001, 64: 053809.

[45] BOCHMANN J, VAINSENCHER A, AWSCHALOM D D, et al. Nanomechanical coupling between microwave and optical photons[J]. Nat. Phys., 2013, 9: 712-716.

[46] FAN L, FONG K Y, POOT M, et al. Cascaded optical transparency in multimode-cavity optomechanical systems[J]. Nat. Commun., 2015,6: 5850.

[47] SOHN D B, KIM S, BAHL G. Time-reversal symmetry breaking with acoustic pumping of nanophotonic circuits[J]. Nat. Photon., 2018, 12: 91-97.

[48] BENDER C M, GIANFREDA M, ÖZDEMIR S K, et al. Twofold transition in PT-symmetric coupled oscillators[J]. Phys. Rev. A, 2013, 88: 062111.

[49] XIONG H, SI L G, ZHENG A S, et al. Higher-order sidebands in optomechanically induced transparency[J]. Phys. Rev. A, 2012, 86: 013815.

[50] DEJESUS E X, KAUFMAN C. Routh-Hurwitz criterion in the examination of eigenvalues of a system of nonlinear ordinary differential equations[J]. Phys. Rev. A, 1987, 35: 5288.

[51] CLERK A A, DEVORET M H, GIRVIN S M, et al. Introduction to quantum noise, measurement, and amplification [J]. Rev. Mod. Phys., 2010, 82: 1155-1208.

[52] VERHAGEN E, DELÉGLISE S, WEIS S, et al. Quantum-coherent coupling of a mechanical oscillator to an optical cavity mode[J]. Nature, 2012, 482: 63-67.

[53] GOLDZAK T, MAILYBAEV A A, MOISEYEV N. Light stops at exceptional points[J]. Phys. Rev. Lett., 2018, 120: 013901.

[54] JIANG C, CUI Y S, ZHAI Z Y, et al. Tunable slow and fast light in parity-time-symmetric optomechanical systems with phonon pump[J]. Opt. Express, 2018, 26(22): 28834-28847.

第 3 章 腔光力学系统中的光学双稳态

3.1 双腔光力学系统中可控的光学双稳态

3.1.1 引言

腔光力学领域研究机械振子与电磁腔通过辐射压力产生的相互作用[1-3]。在过去的十几年中,该领域取得了许多重要的进展,包括机械振子的量子基态冷[4,5]、光力诱导透明[6-8]、光子-声子之间的相干转换[9-11],以及量子态传输腔[12-15]等。单个光子对宏观机械振子施加的辐射压力通常比较小并且本质上是非线性的。目前的实验主要聚焦在强驱动领域,光力耦合强度可以随着腔内光子数的增加而被极大增强[16,17],但是这种增强以失去光子-光子之间相互作用的非线性作为代价。最近,单模[18-24]和双模[25-27]光力系统中有几个理论工作研究了单光子强耦合区域,其中单光子光力耦合强度超过了腔的衰减率。在这个区域,本质上是非线性的光力相互作用在单光子时也比较明显。

光力系统中的光学双稳态是一种重要的非线性效应。最近,包含玻色-爱因斯坦凝聚体(BEC)[28-30]、超冷原子[31-33]和量子阱[34]在内的腔光力学系统中腔内光子数的双稳态行为得到了广泛的研究。由于原子的集体振动,玻色-爱因斯坦凝聚体或者超冷原子中出现双稳行为时腔内的光子数通常比较低,有些甚至低于 1。然而,在空腔组成的典型的光力系统中,通常只有在光子数比较多时才能发生双稳态行为。本研究考虑了两个光学腔共同耦合于一个机械振子构成的双腔光力系统中腔内光子数的双稳态行为,发现通过改变泵浦光束的功率和频率可以有效控制两个腔内光子数的双稳态行为,并且当腔内光子数低于 1 时仍然可以出现双稳态行为,因此在可控的光开关中有着重要的应用。

3.1.2 模型和理论

研究的光力系统如图 3.1.1 所示。两个光学腔模耦合于一个共同的力学模式，它们之间的相互作用哈密顿量为 $H_I = \sum_{k=1,2} \hbar g_k a_k^\dagger a_k (b^\dagger + b)$，其中 a_k 和 b 分别是腔模和力学模式的湮灭算符，g_k 是力学模式和第 k 个腔模之间的单光子耦合强度。物理上，g_k 表示的是机械振子的零点运动引起的第 k 个腔模的频率移动。此外，左边的光学腔同时受到一束强度为 E_L、频率为 ω_L 的强的泵浦场和一束强度为 E_p、频率为 ω_p 的弱的探测场驱动，而右边的光学腔只受到一束强度为 E_R、频率为 ω_R 的强的泵浦场驱动。在泵浦场频率 ω_L 和 ω_R 的旋转框架下，该双腔光力系统的哈密顿表示如下：

$$H = \sum_{k=1,2} \hbar \Delta_k a_k^\dagger a_k + \hbar \omega_m b^\dagger b - \sum_{k=1,2} \hbar g_k a_k^\dagger a_k (b^\dagger + b) + $$
$$i\hbar\sqrt{\kappa_{e,1}} E_L(a_1^\dagger - a_1) + i\hbar\sqrt{\kappa_{e,2}} E_R(a_2^\dagger - a_2) + $$
$$i\hbar\sqrt{\kappa_{e,1}} E_p(a_1^\dagger e^{-i\delta t} - a_1 e^{i\delta t}) \quad (3.1.1)$$

上式右边第一项表示共振频率为 $\omega_k(k=1,2)$ 的腔模的能量，其中 $\Delta_1 = \omega_1 - \omega_L$，$\Delta_2 = \omega_2 - \omega_R$ 分别是腔-泵浦场之间的失谐量。第二项表示共振频率为 ω_m，有效质量为 m 的力学模式的能量。最后三项表示的是输入场与腔场之间的相互作用，其中 E_L、E_R、E_p 与施加激光功率之间的关系分别为 $E_L = \sqrt{2P_L \kappa_1/\hbar \omega_L}$、$E_L = \sqrt{2P_L \kappa_1/\hbar \omega_L}$ 和 $E_L = \sqrt{2P_L \kappa_1/\hbar \omega_L}$（$\kappa_k = \kappa_{i,k} + \kappa_{e,k}$ 表示的是第 k 个腔场的衰减率，其中 $\kappa_{i,k}$ 和 $\kappa_{e,k}$ 分别表示内在衰减率和外部衰减率）。此外，$\delta = \omega_p - \omega_L$ 表示探测场和左泵浦场之间的失谐量。

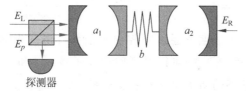

图 3.1.1 两个光学腔 a_1、a_2 耦合于同一个机械振子 b 组成的双腔光力系统原理图。左侧的腔同时受到一束强的泵浦场 E_L 和一束弱的探测场 E_p 驱动，而右侧的腔仅受到一束泵浦场 E_R 的驱动

（请扫Ⅱ页二维码看彩图）

算符 a_1、a_2 和 Q（定义为 $Q = b + b^\dagger$）随时间演化的方程可以根据海森伯运动方程以及对易关系 $[a_k, a_k^\dagger] = 1$ 和 $[b, b^\dagger] = 1$ 得到。引入腔模和力学模

式相应的衰减和噪声项,我们得到以下量子朗之万方程:

$$\dot{a}_1 = -i(\Delta_1 - g_1 Q)a_1 - \kappa_1 a_1 + \sqrt{\kappa_{e,1}}(E_L + E_p e^{-i\delta t}) + \sqrt{2\kappa_1} a_{in,1} \quad (3.1.2)$$

$$\dot{a}_2 = -i(\Delta_2 - g_1 Q)a_2 - \kappa_2 a_2 + \sqrt{\kappa_{e,2}} E_R + \sqrt{2\kappa_2} a_{in,2} \quad (3.1.3)$$

$$\ddot{Q} + \gamma_m \dot{Q} + \omega_m^2 Q = 2g_1 \omega_m a_1^+ a_1 + 2g_2 \omega_m a_2^+ a_2 + \xi \quad (3.1.4)$$

腔模受到平均值为零的输入真空噪声 $a_{in,k}$ 的影响,而衰减率为 γ_m 的力学模式受到平均值为零的布朗随机力 ξ 的影响[35]。

令方程(3.1.2)~方程(3.1.4)对时间求导为零,可以得到以下形式的稳态解:

$$a_{s,1} = \frac{\sqrt{\kappa_{e,1}} E_L}{\kappa_1 + i\Delta_1'}, \quad a_{s,2} = \frac{\sqrt{\kappa_{e,2}} E_R}{\kappa_2 + i\Delta_2'}, \quad Q_s = \frac{2}{\omega_m}(g_1 |a_{s,1}|^2 + g_2 |a_{s,2}|^2)$$
$$(3.1.5)$$

式中,$\Delta_1' = \Delta_1 - g_1 Q_s$,$\Delta_2' = \Delta_2 - g_2 Q_s$ 分别是考虑辐射压效应后腔的有效失谐量。根据劳斯-赫尔维茨(Routh-Hurwitz)判据可以得到系统的稳定性条件[36],形式通常比较复杂。但是,在光力协同性(cooperativity)比较大的极限下,稳定性条件可以近似表达为[37]

$$\widetilde{G}^2 > \bar{C}\gamma_m \max\left[\kappa_1 - \kappa_2, \frac{\kappa_2^2 - \kappa_1^2}{2\gamma_m + \kappa_1 + \kappa_2}\right] \quad (3.1.6)$$

式中,$\widetilde{G} \equiv \sqrt{G_1^2 - G_2^2}$,$\bar{C} \equiv (G_1^2 + G_2^2)/[\gamma_m(\kappa_1 + \kappa_2)]$。腔内平均光子数 $n_{pk} = |a_{s,k}|^2$ 由以下两个方程决定:

$$n_{p1} = \frac{\kappa_{e,1} E_L^2}{\kappa_1^2 + [\Delta_1 - 2g_1/\omega_m(g_1 n_{p1} + g_2 n_{p2})]^2} \quad (3.1.7)$$

$$n_{p2} = \frac{\kappa_{e,2} E_R^2}{\kappa_2^2 + [\Delta_2 - 2g_2/\omega_m(g_1 n_{p1} + g_2 n_{p2})]^2} \quad (3.1.8)$$

这种形式的立方方程表明该系统中腔内光子数可出现光学双稳态[31,32]。从方程(3.1.7)和方程(3.1.8)可以看到腔内光子数 n_{p1} 和 n_{p2} 是相互关联的,可以通过改变泵浦场的功率和频率来改变 E_L、E_R、Δ_1 和 Δ_2,这样可从多方面来控制腔内光子数。例如,可以通过改变右侧泵浦场功率直接控制右侧腔内光子数 n_{p2},也可以通过改变左侧泵浦场功率来间接改变右侧腔内光子数。

3.1.3 结果和讨论

本节将具体的实验参数代入到腔内光子数满足的方程进行数值模拟,从而研究如何对光子数的双稳行为进行调控。用到的参数为[11] $\omega_1 = 2\pi \times 205.3$ THz, $\omega_2 = 2\pi \times 194.1$ THz, $\kappa_1 = 2\pi \times 520$ MHz, $\kappa_2 = 1.73$ GHz, $\kappa_{e,1} = 0.2\kappa_1$, $\kappa_{e,2} = 0.42\kappa_2$, $g_1 = 2\pi \times 960$ kHz, $g_2 = 2\pi \times 430$ kHz, $\omega_m = 2\pi \times 4$ GHz, $Q_m = 87 \times 10^3$,其中 Q_m 是机械振子的品质因数,机械振子的衰减率 γ_m 由 ω_m/Q_m 给出。将 κ_1、κ_2 和 γ_m 的具体数值代入到稳定性条件式(3.1.6),我们发现当 $G_1 > 1.36 G_2$ 时系统是稳定的。以下所取的参数均满足稳定性条件。

此处考虑的双腔光力系统使得腔内光子数的双稳态行为具有更高的可控性。图 3.1.2(a)绘制了左腔内的平均光子数在三个不同泵浦场功率作用下随左腔泵浦失谐量 $\Delta_1 = \omega_1 - \omega_L$ 变化的曲线。当左泵浦的功率为 $P_L = 0.1\ \mu W$ 时,曲线几乎为一个对称的洛伦兹曲线。然而,当功率增加到临界值以上时,系统便出现双稳态现象,如 $P_L = 2\ \mu W$ 和 $P_L = 3\ \mu W$ 的曲线所示,其最初的洛伦兹共振曲线变得不再对称。在这种情况下,腔内平均光子数的耦合三次方程(3.1.7)和方程(3.1.8)产生三个实根。最大和最小的根是稳定的,中间是最不稳定的,由图 3.1.2(a)中的虚线表示。此外,我们可以看到,随着泵浦光束功率的增加,需要更大的腔泵失谐量来观察光学双稳态现象。而腔内平均光子数由图 3.1.2(b)所示的泵功率曲线也可以看出双稳态特性。

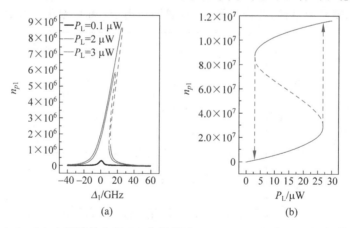

图 3.1.2 (a) 左泵浦场功率 P_L 分别等于 $0.1\ \mu W$、$2\ \mu W$ 和 $3\ \mu W$ 时(从下到上)左腔内的平均光子数随腔泵失谐量 $\Delta_1 = \omega_1 - \omega_L$ 变化的曲线;(b) 左腔内平均光子数随左泵浦场功率 P_L 变化的情况,其中右腔的失谐量为 $\Delta_1 = \Delta_2 = \omega_m$,右侧的泵浦场功率保持为 $0.1\ \mu W$

(请扫Ⅱ页二维码看彩图)

这里,两个腔都被泵浦在各自的红边带,即 $\Delta_1=\Delta_2=\omega_m$。考虑到左泵浦场功率逐渐从零增加;腔内平均光子数 n_1 最初位于稳定分支(对应于最小根)。当泵浦场功率 P_L 增加到临界值时(约为 27 μW),n_1 接近该曲线的末端。如果 P_L 进一步增加,则 n_1 跳跃到上分支并继续增加。如果 P_L 减小,则腔内光子数沿着上支不断减小直到临界值。随着 P_L 进一步减小,则光子数又跃迁到较低的稳定分支。

接下来我们主要通过控制左泵浦光束的频率和功率来研究右腔中的光学双稳行为。图 3.1.3 所示为以右腔内的平均光子数 n_2 随左腔泵浦失谐量 Δ_1 为变化的曲线。当左腔和机械振子之间的耦合关闭时,即 $g_1=0$ 时,双腔光力系统成为普通的单模腔光力学系统,在这种情况下左侧腔内泵浦驱动不会影响右侧腔内光子数。此时,如果对右腔的泵浦场施加红失谐驱动,则从图 3.1.3 的中间部分可以清楚地看出,当左腔泵失谐量 Δ_1 改变时,腔内平均光子数 n_2 保持恒定值。然而,如果左腔和机械振子之间的耦合打开,则右腔中的腔内平均光子数的双稳态行为将出现。当 $\Delta_2=\omega_m$ 时,平均光子数大于之前的恒定值。然而,如果对右腔的泵浦场施加蓝失谐驱动,即 $\Delta_2=-\omega_m$,则平均光子数小于上述常数值。因为当 $g_1=0$ 和 $\Delta_2=\omega_m$ 时,混合系统转向普通的单模光力系统,腔内光子数、泵浦场功率 P_R 和腔泵失谐量 Δ_2^2 的平方

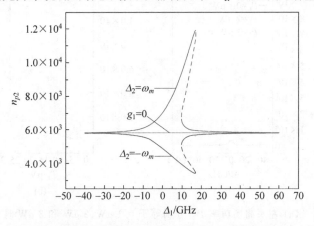

图 3.1.3 右腔内的平均光子数随左腔泵浦失谐量 Δ_1 变化的曲线,而右腔泵浦失谐量分别输入为 $\Delta_2=\omega_m$ 和 $\Delta_2=-\omega_m$。左泵浦场功率 $P_L=2$ μW,右泵浦场功率 $P_R=0.1$ μW

(请扫 Ⅱ 页二维码看彩图)

直接相关。因此,当 $g_1=0$,$P_R=0.1~\mu W$,$\Delta_2=\pm\omega_m$ 时,光子数保持不变。然而,当 $g_1\neq 0$ 时,左腔中的腔内光子将对共同的机械振子以及右腔中的光子数有影响。当 $\Delta_2=\omega_m$ 时,频率 $\omega_R-\omega_m$ 处的高度非共振的斯托克斯散射被强烈抑制,此时仅存在于右腔内积聚在频率 $\omega_R+\omega_m$ 处的反斯托克斯散射会导致泵上的光子以频率 ω_2 转换到腔光子。因此,右腔中的平均光子数大于恒定值,而不受左腔的影响。所以,通过调节左腔泵浦光束失谐量 Δ_1,可以观察到右腔内的腔内光子数的双稳态。如图 3.1.4 所示,当 $\Delta_2=\omega_m$ 和 $\Delta_2=-\omega_m$ 时,右腔中的光学双稳态也可以从腔内平均光子数与左泵浦场功率的滞后回线中看出。在这里,我们将左腔泵浦束失谐量定为 $\Delta_1=\omega_m$ 和 $P_R=0.1~\mu W$。类似地,当 $\Delta_2=\omega_m$ 时的腔内平均光子数大于当 $\Delta_2=-\omega_m$ 时腔内的光子数。

在之前的讨论中,我们已经证实了两个腔中的光学双稳态,并且腔内平均光子数通常非常大,在右腔中至少有数千个光子(图 3.1.3 和图 3.1.4)。单模光力系统($g_2=0$)将需要更多的光子在空腔中以达到双稳态(图 3.1.2)。接下来,我们将展示双腔光力系统在极低的腔光子数下实现光学双稳态。

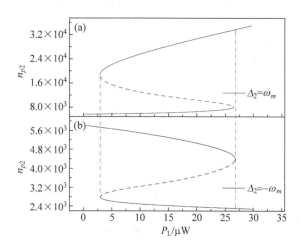

图 3.1.4 右腔内的平均光子数随左泵浦场功率变化的曲线,其中 $P_R=0.1~\mu W$,
(a)$\Delta_2=\omega_m$ 而(b)$\Delta_2=-\omega_m$。左腔泵失谐量 Δ_1 始终等于 ω_m
(请扫 II 页二维码看彩图)

图 3.1.5(a)和(b)所示分别是右腔中的平均光子数随左腔泵浦失谐量 Δ_1 及左腔泵浦场功率 P_L 变化的曲线。这里右泵浦光束的参数为 $P_R=1~pW$,$\Delta_2=\omega_m$。由于泵浦场功率较低,右腔中的腔内光子数非常小,即 $n_2\leqslant 1$。通

常,这种低光子数不能在空腔光力系统中出现双稳现象。然而,在这里考虑的双腔光力系统中,当左腔由强泵浦场驱动时,由于两个腔耦合到共同的纳米机械振子,光学双稳态仍然存在于右腔中。这个行为可以理解如下:由左腔施加的辐射压力引起机械振子的振动,这改变了两个空腔的光路长度,从而产生了腔场上的位置相移。低光子数的双稳态是由光子和声子之间的这种非线性反馈产生的。这种现象表示弱耦合状态下的强非线性效应,这是通过力学模式的长寿命和左腔上的强泵实现的。最近,吕(Lu)等[38]和库扎克(Kuzyk)等[39]的两个相关工作还表明,在弱耦合状态下,双腔光力系统可以获得强非线性。此外,双腔光力系统中腔内光子数的双稳行为可以用作一种可控的光开关,而右腔中光子数的两个稳定分支起着光学开关的作用。当左侧泵浦场的频率和功率固定时,通过控制右泵浦波束的频率和功率,可以很容易实现较低稳态的分支与较高稳态分支之间的切换。此外,左侧泵浦光束可用作控制参数以打开或关闭该开关。

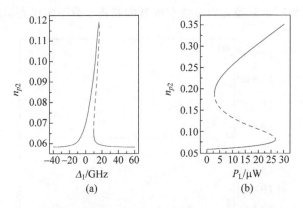

图3.1.5 右腔内的平均光子数随(a)左腔泵浦失谐量 Δ_1 变化而 $P_L = 2\ \mu W$;(b)左泵浦场功率 P_L 变化而 $\Delta_1 = \omega_m$ 的曲线。其他参数为 $P_R = 1\ pW, \Delta_2 = \omega_m$

(请扫Ⅱ页二维码看彩图)

3.1.4 小结

本节研究了由两个光学腔通过辐射压力耦合于一个共同的机械振子形成的双腔光力系统中的光学双稳现象[40]。与通常的单模腔光力学系统相比,这里的双腔光力系统可以更加灵活地控制光学双稳态。其中一个腔中的平均腔内光子数可以通过另外一个腔的泵浦场的功率和频率进行控制。此外，在这个耦合系统中光学双稳态在腔内光子数非常低时依然可能存在。

参 考 文 献

[1] KIPPENBERG T J, VAHALA K J. Cavity optomechanics: back-action at the mesoscale[J]. Science,2008,321: 1172.

[2] MARQUARDT F,GIRVIN S M. Optomechanics[J]. Physics,2009,2: 40.

[3] ASPELMEYER M, KIPPENBERG T J, MARQUARDT F. Cavity optomechanics [J]. Rev. Mod. Phys. ,2014,86: 1391.

[4] TEUFEL J D,DONNER T,LI D,et al. Sideband cooling of micromechanical motion to the quantum ground state[J]. Nature,2011,475: 359.

[5] CHAN J, ALEGRE T P, SAFAVI-NAEINI A H, et al. Sideband cooling of micromechanical motion to the quantum ground state[J]. Nature,2011,478: 89.

[6] AGARWAL G S, HUANG S. Electromagnetically induced transparency in mechanical effects of light[J]. Phys. Rev. A,2010,81: 041803.

[7] WEIS S,RIVIÈRE R,DELÉGLISE S,et al. Optomechanically induced transparency [J]. Science,2010,330: 1520.

[8] SAFAVI-NAEINI A H, ALEGRE T P M, CHAN J, et al. Electromagnetically induced transparency and slow light with optomechanics[J]. Nature,2011,472: 69.

[9] FIORE V,YANG Y,KUZYK M C,et al. Storing optical information as a mechanical excitation in a silica optomechanical resonator [J]. Phys. Rev. Lett. , 2011, 107: 133601.

[10] VERHAGEN E, DELÉGLISE S, WEIS S, et al. Quantum-coherent coupling of a mechanical oscillator to an optical cavity mode[J]. Nature,2012,482: 63.

[11] HILL J T, SAFAVI-NAEINI A H, CHAN J, et al. Coherent optical wavelength conversion via cavity optomechanics[J]. Nat. Commun. ,2012,3: 1196.

[12] TIAN L, WANG H L. Optical wavelength conversion of quantum states with optomechanics[J]. Phys. Rev. A,2010,82: 053806.

[13] WANG Y D, CLERK A A. Using interference for high fidelity quantum state transfer in optomechanics[J]. Phys. Rev. Lett. ,2012,108: 153603.

[14] TIAN L. Adiabatic state conversion and pulse transmission in optomechanical systems[J]. Phys. Rev. Lett. ,2012,108: 153604.

[15] PALOMAKI T A, HARLOW J W, TEUFEL J D,et al. Coherent state transfer between itinerant microwave fields and a mechanical oscillator[J]. Nature,2013, 495: 210.

[16] GRÖBLACHER S,HAMMERER K,VANNER M R,et al. Observation of strong

coupling between a micromechanical resonator and an optical cavity field[J]. Nature,2009,460: 724.

[17] TEUFEL J D,LI D,ALLMAN M S,et al. Circuit cavity electromechanics in the strong-coupling regime[J]. Nature,2011,471: 204.

[18] RABL P. Photon blockade effect in optomechanical systems[J]. Phys. Rev. Lett., 2011,107: 063601.

[19] NUNNENKAMP A,BØRKJE K,GIRVIN S M. Single-photon optomechanics[J]. Phys. Rev. Lett. ,2011,107: 063602.

[20] LIAO J Q,CHEUNG H K,LAW C K. Spectrum of single-photon emission and scattering in cavity optomechanics[J]. Phys. Rev. A,2012,85: 025803.

[21] BØRKJE K,NUNNENKAMP A,TEUFEL J D,et al. Signatures of nonlinear cavity optomechanics in the weak coupling regime [J]. Phys. Rev. Lett., 2013, 111: 053603.

[22] LEMONDE M A,DIDIER N,CLERK A A. Nonlinear interaction effects in a strongly driven optomechanical cavity[J]. Phys. Rev. Lett.,2013,111: 053602.

[23] KRONWALD A,MARQUARDT F. Optomechanically induced transparency in the nonlinear quantum regime[J]. Phys. Rev. Lett.,2013,111: 133601.

[24] HE B. Quantum optomechanics beyond linearization[J]. Phys. Rev. A, 2012, 85: 063820.

[25] LUDWIG M, SAFAVI-NAEINI A H, PAINTER O, et al. Enhanced quantum nonlinearities in a two-mode optomechanical system[J]. Phys. Rev. Lett. ,2012, 109: 063601.

[26] STANNIGEL K,KOMAR P,HABRAKEN S J M,et al. Optomechanical quantum information processing with photons and phonons[J]. Phys. Rev. Lett. ,2012,109: 013603.

[27] KÓMÁR P,BENNETT S D,STANNIGEL K,et al. Single-photon nonlinearities in two-mode optomechanics[J]. Phys. Rev. A,2013,87: 013839.

[28] BRENNECKE F,RITTER S,DONNER T,et al. Cavity optomechanics with a Bose-Einstein condensate[J]. Science,2008,322: 235.

[29] ZHANG J M,CUI F C,ZHOU D L,et al. Nonlinear dynamics of a cigar-shaped Bose-Einstein condensate in an optical cavity[J]. Phys. Rev. A,2009,79: 033401.

[30] YANG S,AL-AMRI M,ZUBAIRY M S. Anomalous switching of optical bistability in a Bose-Einstein condensate[J]. Phys. Rev. A,2013,87: 033836.

[31] GUPTA S, MOORE K L, MURCH K W, et al. Cavity nonlinear optics at low photon numbers from collective atomic motion [J]. Phys. Rev. Lett., 2007, 99: 213601.

[32] KANAMOTO R,MEYSTRE P. Optomechanics of a quantum-degenerate Fermi gas[J]. Phys. Rev. Lett.,2010,104: 063601.

[33] PURDY T P,BROOKS D W C,BOTTER T,et al. Tunable cavity optomechanics

with ultracold atoms[J]. Phys. Rev. Lett. ,2010,105: 133602.
[34] SETE E A, ELEUCH H. Controllable nonlinear effects in an optomechanical resonator containing a quantum well[J]. Phys. Rev. A,2012,85: 043824.
[35] GENES C, VITALI D, TOMBESI P, et al. Ground-state cooling of a micromechanical oscillator: comparing cold damping and cavity-assisted cooling schemes[J]. Phys. Rev. A,2008,77: 033804.
[36] DEJESUS E X, KAUFMAN C. Routh-Hurwitz criterion in the examination of eigenvalues of a system of nonlinear ordinary differential equations[J]. Phys. Rev. A,1987,35: 5288.
[37] WANG Y D,CLERK A A. Reservoir-engineered entanglement in optomechanical systems[J]. Phys. Rev. Lett. ,2013,110: 253601.
[38] LÜ X Y, ZHANG W M, ASHHAB S, et al. Quantum-criticality-induced strong Kerr nonlinearities in optomechanical systems[J]. Sci. Rep. ,2013,3: 2943.
[39] KUZYK M C,ENK S J,WANG H L. Generating robust optical entanglement in weak-coupling optomechanical systems[J]. Phys. Rev. A,2013,88: 062341.
[40] JIANG C,LIU H X,CUI Y S,et al. Controllable optical bistability based on photons and phonons in a two-mode optomechanical system[J]. Phys. Rev. A, 2013, 88: 055801.

3.2 二能级原子与腔场耦合的混杂光力系统中的光学双稳态和动力学效应

3.2.1 引言

腔光力学领域研究通过辐射压力产生的光与机械运动之间的非线性相互作用[1-3]。单光子耦合常数一般情况下较弱，但是相干地驱动腔场可以极大提高有效光力相互作用。在过去的十年里，像机械振子的基态冷却[4,5]、光力诱导透明[6-8]以及光学双稳态[9,10]等这样一些有意思的现象已经在驱动腔光力系统中被发现。然而，在强驱动情况下的光力相互作用变成线性的。因此，路德维希(Ludwig)和苏埃雷布(Xuereb)等作了相当大的努力来提高内在的非线性[11-16]。最近增加强非线性，例如一个量子两级体系，使得光力系统中的物理效应更加丰富[17-23]。谐振腔通过辐射压与一个机械振子耦合，并且通过J-C与一个二能级原子耦合，因此混杂系统结合了腔量子电动力学和腔光力学。假设原子始终处于它的激发态，王(Wang)等首次理论性地研究了透明和放大现象[24],之后阿克拉姆(Akram)等研究了在混杂光力系统中可调的快慢光效应[25]。

在先前的研究中,在含有一个量子阱[26]、超冷原子[27-29]以及一个玻色-爱因斯坦凝聚体[30-33]的光力系统中平均腔内光子数的双稳态行为已经受到了广泛的研究。近年来,驱动空腔的光场和另一个直接与玻色-爱因斯坦凝聚体相互作用的横向场已经被运用在混杂玻色-爱因斯坦凝聚体光力系统中控制双稳态[34]。此外,在耦合里德伯超级原子的光力系统中镜子位移、腔光子数以及里德伯原子布居数的双稳态特性受到了研究[35]。熊(Xiong)等研究了交叉克尔效应对平均光子数稳态行为的影响[36],而交叉克尔效应可以通过将一个二能级系统同时耦合到腔模和机械振子实现[18,19,22]。

基于以上的研究进展,本节研究含有二能级原子的混杂腔光力学系统中的光学双稳态和动力学效应。特别指出,本书考虑的混杂系统和雷斯特雷波(Restrepo)等提出的系统相同[17,23-25],在此系统中二能级原子通过 J-C 耦合被耦合到腔场。如果再考虑二能级原子与机械振子的耦合[37],例如赫拉(Heikkila)等研究腔与力学模式之间的交叉克尔型耦合[18,22],其可以引起平均光子数的双稳态和三稳态行为[36]。我们采用和亚西尔(Yasir)等相似的方法[34]对系统中的光学双稳态和动力学效应等进行研究,将一个强泵浦场施加到腔场上,而另一个横向驱动场施加到原子上。研究发现腔泵浦场和原子驱动场可被用来控制稳态腔光子数和原子布居数的双稳行为。当腔泵浦场强度相对较小时,原子引入的非线性会引起一个附加的双稳区域,并且腔-原子之间的耦合强度越大,双稳区域越宽。此外,我们对不同初始条件和腔泵浦场功率对系统的动力学效应进行了分析。与文献[34]中考虑的包含玻色-爱因斯坦凝聚体的混杂光力系统比较而言,这里讨论的混杂光力系统有着一些明显的优势。首先,包含二能级原子(量子比特)的光力系统将腔量子电动力学和腔光力学结合起来,其中腔-原子之间的相互作用实现一种非经典的光子探测器,进而可以提供一种控制长寿命机械量子态的非线性源。但是混杂玻色-爱因斯坦腔光力学系统中玻色-爱因斯坦凝聚体的集体密度激发起到机械振子的作用,所以单个原子的非线性动力影响变得很弱。尤其是,包含单个原子的混杂光力系统中可以出现两个不同的双稳区域,但是在包含玻色-爱因斯坦凝聚体的光力系统中只有一个双稳区域。其次,在使用人造原子实验实现的光力系统中[23],腔-量子比特之间的相互作用可以通过对量子比特的失谐量进行有效调节,这使得我们能够更加灵活地控制系统的双稳态行为和动力学效应。

3.2.2 模型和理论

我们考虑一个在腔内包含二能级原子的混杂光力系统,如图 3.2.1 所示。腔模通过 J-C 耦合形式耦合到二能级原子,并通过辐射压力耦合到机械振子。我们假设将频率为 ω_p 的强泵浦场施加到腔上,并且将频率为 ω_d 的强驱动场施加到两能级原子上。系统的总哈密顿量可以写成

$$\hat{H} = \hbar \omega_c \hat{a}^+ \hat{a} - \hbar G \hat{a}^+ \hat{a} \hat{q} + \frac{1}{2} \hbar \omega_{at} \hat{\sigma}_z + \frac{\hat{p}^2}{2m} + \frac{1}{2} m \omega_m^2 \hat{q}^2 + $$
$$\hbar J(\hat{a} \hat{\sigma}_+ + \hat{a}^+ \hat{\sigma}_-) + i \hbar E_p (\hat{a}^+ e^{-i\omega_p t} - \hat{a} e^{i\omega_p t}) + \mu E_d (\hat{\sigma}_+ e^{-i\omega_d t} + \hat{\sigma}_- e^{i\omega_d t})$$
(3.2.1)

其中,ω_c 是具有产生(湮没)算符 $\hat{a}^+(\hat{a})$ 的谐振腔模式的共振频率,ω_{at} 是态 $|g\rangle$ 与原子激发态 $|e\rangle$ 之间的跃迁频率,$\hat{\sigma}_z \equiv |e\rangle\langle e| - |g\rangle\langle g|$ 是通常所说的泡

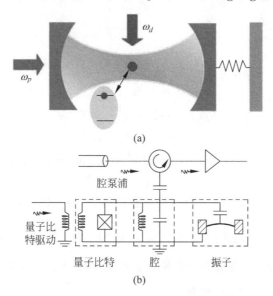

图 3.2.1 混杂光力系统示意图

(a) 一个二能级原子被放置在一个光机械腔内,它包括固定镜和可移动镜。谐振腔受到频率为 ω_p 的强泵浦场驱动,频率为 ω_d 的驱动场作用于二能级原子;(b) 等效电路图

(请扫Ⅱ页二维码看彩图)

利算符,$\hat{\sigma}_+$($\hat{\sigma}_-$)是原子上升(下降)算符,$\hat{q}$ 和 $\hat{p}$ 表示振动频率为 ω_m 和质量为 m 的机械振子的位置和动量算符。方程(3.2.1)右边的第二项描述了腔场与机械振子之间的耦合,其中 $G=\mathrm{d}\omega_c/\mathrm{d}q$ 表示腔谐振频率随机械振子位移的变化。第六项代表腔场与二能级原子之间的相互作用,J 为耦合强度。最后两项表示外加的场与混杂系统之间的相互作用,其中 E_p 是腔泵浦场的振幅,μ 是原子的电偶极矩,E_d 是原子驱动场的振幅。在腔泵浦场的频率为 ω_p 的旋转框架中,方程(3.2.1)变成

$$\hat{H} = \hbar\Delta_c\hat{a}^+\hat{a} - \hbar G\hat{a}^+\hat{a}\hat{q} + \frac{1}{2}\hbar\Delta_{at}\hat{\sigma}_z + \frac{\hat{p}^2}{2m} + \frac{1}{2}m\omega_m^2\hat{q}^2 +$$
$$\hbar J(\hat{a}\hat{\sigma}_+ + \hat{a}^+\hat{\sigma}_-) + \mathrm{i}\hbar E_p(\hat{a}^+ - \hat{a}) + \hbar\Omega_d(\hat{\sigma}_+ + \hat{\sigma}_-) \quad (3.2.2)$$

式中,$\Delta_c = \omega_c - \omega_p$ 是腔泵浦场失谐量,$\Delta_{at} = \omega_{at} - \omega_p$ 是原子驱动场失谐量,我们假设 $\omega_p = \omega_d$,$\Omega_d = \mu E_d/\hbar$ 是原子驱动场的拉比频率。

系统的量子朗之万方程可以通过运用海森伯运动方程并加入相应的阻尼和输入噪声项:

$$\frac{\mathrm{d}\hat{q}}{\mathrm{d}t} = \frac{\hat{p}}{m} \quad (3.2.3)$$

$$\frac{\mathrm{d}\hat{p}}{\mathrm{d}t} = -m\omega_m^2\hat{q} - \gamma_m\hat{p} + \hbar G\hat{a}^+\hat{a} + \hat{\xi}(t) \quad (3.2.4)$$

$$\frac{\mathrm{d}\hat{a}}{\mathrm{d}t} = -(\kappa + \mathrm{i}\Delta_c)\hat{a} + \mathrm{i}G\hat{a}\hat{q} - \mathrm{i}J\hat{\sigma}_- + E_p + \sqrt{2\kappa}\hat{a}_{in}(t) \quad (3.2.5)$$

$$\frac{\mathrm{d}\hat{\sigma}_-}{\mathrm{d}t} = -(\mathrm{i}\Delta_{at} + \Gamma_2)\hat{\sigma}_- + \mathrm{i}J\hat{a}\hat{\sigma}_z + \mathrm{i}\Omega_d\hat{\sigma}_z + \sqrt{2\Gamma_2}\hat{c}_{in}(t) \quad (3.2.6)$$

$$\frac{\mathrm{d}\hat{\sigma}_z}{\mathrm{d}t} = -\Gamma_1(\hat{\sigma}_z + 1) - 2\mathrm{i}J(\hat{a}\hat{\sigma}_+ - \hat{a}^\dagger\hat{\sigma}_-) - 2\mathrm{i}\Omega_d(\hat{\sigma}_+ - \hat{\sigma}_-) \quad (3.2.7)$$

式中,κ、γ_m 和 Γ_1 分别是腔场、机械振子和二能级原子的衰减率,Γ_2 是原子的退相干率。腔场和二能级原子受输入真空噪声影响,$\hat{a}_{in}(t)$ 和 $\hat{c}_{in}(t)$ 的平均值为零且在时域中遵循关联函数:

$$\hat{a}_{in}(t)\hat{a}_{in}^+(t') = \delta(t-t') \quad (3.2.8)$$

$$\hat{c}_{in}(t)\hat{c}_{in}^+(t') = \delta(t-t') \quad (3.2.9)$$

机械阵子受布朗随机力 $\xi(t)$ 影响,其平均值为零且满足关联函数:

$$\langle \xi(t)\xi(t')\rangle = \frac{\gamma_m}{\omega_m}\int\frac{d\omega}{2\pi}\omega e^{-i\omega(t-t')}\left[1+\coth\left(\frac{\hbar\omega}{2k_B T}\right)\right] \quad (3.2.10)$$

式中 k_B 是玻耳兹曼常数，T 是机械振子的库的温度。在这项工作中，我们感兴趣的是平均响应，即经典的运动方程，所以运算符可以用它们的期望值来代替，即 $Q(t)\equiv\langle\hat{Q}(t)\rangle$（其中 $Q\in\{q,p,a,\sigma_-,\sigma_z\}$）。通过对平均值进行因式分解来使用平均场近似，即 $\langle Q_c\rangle\equiv\langle Q\rangle\langle c\rangle$，可以得到以下期望值方程：

$$\frac{dq}{dt}=\frac{p}{m} \quad (3.2.11)$$

$$\frac{dp}{dt}=-m\omega_m^2 q-\gamma_m p+\hbar G a^* a \quad (3.2.12)$$

$$\frac{da}{dt}=-(\kappa+i\Delta_c)a+iGaq-iJ\sigma_-+E_p \quad (3.2.13)$$

$$\frac{d\sigma_-}{dt}=-(i\Delta_{at}+\Gamma_2)\sigma_-+iJa\sigma_z+i\Omega_d\sigma_z \quad (3.2.14)$$

$$\frac{d\sigma_z}{dt}=-\Gamma_1(\sigma_z+1)-2iJ(a\sigma_+-a^*\sigma_-)-2i\Omega_d(\sigma_+-\sigma_-) \quad (3.2.15)$$

式(3.2.11)~式(3.2.15)的稳态解可以通过将所有时间导数设置为零容易地导出，这由下式给出：

$$q_s=\frac{\hbar G|a_s|^2}{m\omega_m^2} \quad (3.2.16)$$

$$p_s=0 \quad (3.2.17)$$

$$a_s=\frac{E_p-iJL_s}{\kappa+i(\Delta_c-Gq_s)} \quad (3.2.18)$$

$$L_s=\frac{i(Ja_s+\Omega_d)}{\Gamma_2+i\Delta_{at}}W_s \quad (3.2.19)$$

$$\Gamma_1(W_s+1)=-2iJ(a_s L_s^*-a_s^* L_s)-2i\Omega_d(L_s^*-L_s) \quad (3.2.20)$$

其中我们将 L_s 和 W_s 分别设为 σ_- 和 σ_z 的稳态解。需要注意的是，只有当与方程(3.2.11)~方程(3.2.15)相关的线性化朗之万方程的所有本征值的实部都为负时系统才是稳定的。稳定性条件可以通过应用 Routh-Hurwitz 判据[38]来推导，但其一般形式太过复杂，这里不再给出。但是，我们数值验证了本书中选择的参数满足稳定性条件。根据式(3.2.16)~式(3.2.20)，腔内光子数 $n_p=|a_s|^2$ 和粒子数反转 W_s 的稳态解由以下耦合方程确定：

$$n_p[(\kappa - \tilde{J}\Gamma_2 W_s)^2 + (\tilde{\Delta}_c + \tilde{J}\Delta_{at} W_s)^2] = (E_p + \tilde{J}\tilde{\Omega}\Gamma_2 W_s)^2 + \tilde{J}^2\tilde{\Omega}^2\Delta_{at}^2 W_s^2 \quad (3.2.21)$$

$$\Gamma_1(W_s + 1)[(\kappa\Delta_{at} + \Gamma_2\tilde{\Delta}_c)^2 + (\Gamma_2\kappa - \Delta_{at}\tilde{\Delta}_c - J^2 W_s)^2] +$$
$$4\Gamma_2 W_2[\Omega_d^2(\kappa^2 + \tilde{\Delta}_c^2) + \cdots + E_p J^2(E_p + 2\tilde{\Omega}\kappa)] = 0 \quad (3.2.22)$$

式中，

$$\tilde{J} = \frac{J^2}{\Gamma_2^2 + \Delta_{at}^2}, \quad \tilde{\Omega} = \frac{\Omega_d}{J}, \quad \tilde{\Delta}_c = \Delta_c - \frac{\hbar G^2 n_p}{m\omega_m^2} \quad (3.2.23)$$

这种形式的耦合三次方程式具有光学多稳态性的特征[27,29]，而且可以通过拉比频率 Ω_d、泵浦强度 E_p、腔泵失谐 Δ_c 和腔-原子耦合强度 J 来控制。

3.2.3 光子数和布居数反转的双稳态行为

本节我们将选择实际参数来数值上研究腔内光子数和粒子反转的稳态解。使用的参数如下[23]：$\omega_c = 2\pi \times 10.188 \text{ GHz}, \kappa = 2\pi \times 163 \text{ kHz}, \omega_m = 2\pi \times 15.9 \text{ MHz}, \gamma_m = 2\pi \times 150 \text{ Hz}, G = 2\pi \times 95 \text{ MHz/nm}$，机械零点运动 $q_{zpf} = 3.18 \text{ fm}, J = 2\pi \times 12.5 \text{ MHz}, \Gamma_2 = 1/170 \text{ GHz}, \Gamma_1 = 2\Gamma_2$。我们假设量子比特最初处于基态，因此 $W_s = -1$ 为初始值。在这个混杂系统中，腔场同时耦合到力学模式和二能级原子。从上述参数可以看出，单光子光机械耦合率 $g_0 = Gq_{zpf}$ 远小于腔-原子耦合常数 J。但是，如果外场驱动腔和原子，可以增加腔内光子数 n_p；因此，有效光机械耦合强度 $g = g_0\sqrt{n_p}$ 变大。从式(3.2.21)和式(3.2.22)可以看出，泵浦率和拉比频率可以用来控制平均腔内光子数和原子的粒子数反转的双稳态行为。

平均腔内光子数 n_p 和原子的粒子数反转 W_s 随归一化拉比频率 Ω_d/Γ_1 变化的情况如图3.2.2所示，其中 $E_p/2\pi = 0$、0.5 MHz、1 MHz。当腔泵浦场关闭时，归一化的拉比频率增加到临界值以上，耦合方程(3.2.21)和方程(3.2.22)中的平均腔内光子数和粒子数反转有三个实根。由图3.2.2(a)和(b)中的虚线可以看出，最大和最小的根是稳定的，中间的是不稳定的。因此，原子驱动场可以用来控制混杂系统的双稳态行为。此外，从 $E_p = 2\pi \times 0.5$ 和 1 MHz 曲线可以看出，随着腔泵浦强度的增加，观察双稳态行为所需要的拉比频率的临界值变小。

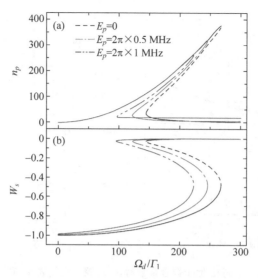

图 3.2.2 $E_p/(2\pi)=0$、0.5 MHz、1 MHz 时(a)平均腔内光子数 n_p、(b)稳态粒子数反转 W_s 与归一化拉比频率 Ω_d/Γ_1 的关系曲线

(请扫Ⅱ页二维码看彩图)

在没有单原子的通常的光力系统中,平均腔内光子数的光学双稳态主要受腔泵浦场控制。在下文中,我们将重点研究在没有原子驱动场的情况下系统的双稳态行为,即 $\Omega_d=0$。图 3.2.3 所示是不同的腔-原子耦合强度时平均内腔光子数 n_p 和粒子数反转 W_p 随腔泵浦场的振幅 E_p 变化的情况。当腔与原子之间没有直接耦合时,从图 3.2.3(a)中的实线可以看出,腔内光子数随腔泵浦强度的增加而单调增加。如图 3.2.3(b)和(d)所示,当腔泵浦场的振幅 E_p 变化时,原子保持基态($W_s=-1$)。然而,如果腔-原子耦合被打开,我们可以看到当腔泵浦强度 $E_p/(2\pi)$ 增加到一个小临界值以上时,出现低腔内光子数和原子布居数反转的双稳态行为。腔-原子耦合常数 J 越大,双稳态行为的范围就越宽,这是由于原子引起的非线性增加所致。当腔泵浦强度 $E_p/(2\pi)$ 进一步升高至几兆赫兹,如图 3.2.3(c)和(d)所示,由于有效光力耦合强度的增加,不同腔-原子耦合对腔内光子数的光学双稳态的影响几乎可以忽略不计,但在强的泵浦场作用下,粒子数反转 $W_s\approx 0$。因此,在这种混合光力系统中,腔内光子数的双稳态行为可以出现在泵浦场的两个不同频率范围内,但是当腔泵浦强度增加到临界值以上时,粒子数反转将总是近似等于零,这意味着原子占据激发态和基态的概率相同。

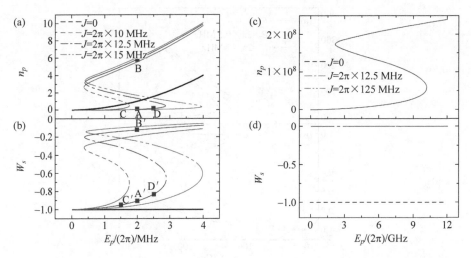

图 3.2.3 (a)和(c)不同腔-原子耦合常数时,平均腔内光子数 n_p 和(b)与(d)稳态布居数反转 W_s 随泵浦率 $E_p/(2\pi)$ 变化的曲线

(请扫Ⅱ页二维码看彩图)

双稳态行为也可以从 n_p 和 W_s 稳态解的滞后回线与腔-泵浦场失谐量 $\Delta_c/(2\pi)$ 的关系中看出。图 3.2.4(a)显示,当腔泵浦场的振幅 $E_p=2\pi\times 0.4\,\text{GHz}$ 时,曲线几乎为洛伦兹型;因此固定腔泵浦失谐的平均腔内光子数只有一个稳定的实解。然而,当腔泵浦场的振幅增加到临界值以上时,平均光子数可以表现出双稳态行为,如 $E_p=2\pi\times 1\,\text{GHz}$、$1.5\,\text{GHz}$、$2\,\text{GHz}$ 的曲线所示,其中最初的洛伦兹共振曲线变得不对称。在如图 3.2.4(a)中还可以看到,随着腔泵浦场振幅的增加,需要更大的腔泵浦失谐量 Δ_c 才能观察到双稳态行为。此外,图 3.2.4(b)绘制了腔泵浦场的振幅 $E_p=2\pi\times 0.1\,\text{MHz}$、$2\,\text{MHz}$、$3\,\text{MHz}$、$4\,\text{MHz}$ 时布居数反转 W_s 随腔-泵浦失谐量 Δ_c 变化的情况。当腔泵浦场的振幅 $E_p=2\pi\times 0.1\,\text{MHz}$ 时,腔泵浦场对量子比特的影响较弱,粒子数反转 W_s 几乎保持不变。随着腔泵浦强度的增加,腔泵浦失谐量发生变化时会出现粒子数反转的双稳态行为。基于以上讨论,可以得出结论,平均腔内光子数的双稳态行为和粒子数反转可以通过腔泵浦场的频率和振幅以及原子驱动场的拉比频率来有效控制。

3.2.4 初始条件和腔泵浦强度对系统动力学效应的影响

上述研究表明,稳态腔内光子数 n_p 和布居数反转 W_s 可以出现双稳态行

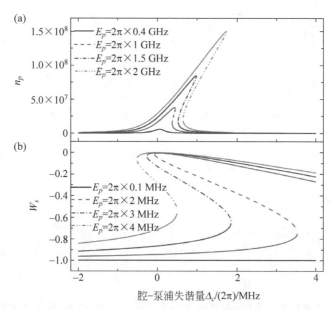

图 3.2.4 (a)腔内光子平均数 n_p 和(b)稳态粒子数反转 W_s 随腔泵浦失谐量 Δ_c 变化的曲线

(请扫Ⅱ页二维码看彩图)

为,本节主要研究不同初始条件和泵浦率时腔内光子数和布居数反转随时间演化的情况。在长时极限下,腔内光子数和粒子数反转可以分别达到它们的稳态解 n_p 和 W_s,它们应该与上面所示的双稳态曲线中相应的点对应。我们采用龙格-库塔法求解常微分方程(3.2.11)~方程(3.2.15),选取 $q|_{t=0}=0$, $p|_{t=0}=0, \sigma_-|_{t=0}=0, \sigma_z|_{t=0}=-1$ 作为初始条件。

图 3.2.5 绘制了 $E_p=2\pi\times 2$ MHz 和 $J=2\pi\times 12.5$ MHz 时腔内光子数 $|a|^2$ 和布居数反转 σ_z 在不同初始条件下随时间的演化曲线。腔场的初始状态可以通过将腔泵浦强度调整到某个初始值来控制,然后当腔泵浦强度突然调整到 $E_p/(2\pi)=2$ MHz 时,系统将达到新的稳态。从图 3.2.5(a)和(b)可以看出,当 $a|_{t=0}=0$ 和 $a|_{t=0}=0.5$ 时,腔内光子数 $|a|^2$ 和布居数反转 σ_z 达到它们的稳态解 $|a_s|^2\approx 0.084, W_s\approx -0.906$,它们对应于图 3.2.3(a)中的 A 点(2,0.084)和图 3.2.3(b)中 A′点(2,-0.906)。但是,如果腔场的初始条件变为 $a|_{t=0}=1$ 和 $a|_{t=0}=2$ 时,$|a|^2$ 和 σ_z 的稳态解将变为 5.857 和 0.122, 分别对应于图 3.2.3(a)中的点 B(2,5.857)和图 3.2.3(b)中的点 B′(2,0.122)。

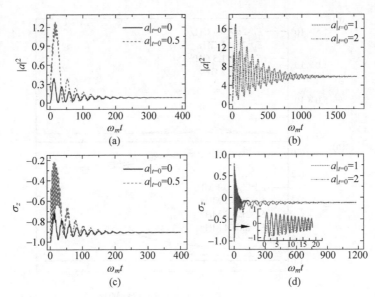

图 3.2.5　不同初始条件下,(a)和(c)腔内光子数$|a|^2$与(b)和(d)布居数反转σ_z随归一化时间$\omega_m t$演化的曲线

(请扫Ⅱ页二维码看彩图)

因此,腔内光子数和布居数反转的稳态解可以对应双稳态曲线的下分支或上分支中的稳态,这主要依赖于腔场的初始值。

此外,图 3.2.3 表明腔内光子数和布居数反转可以通过增加腔泵浦场的振幅E_p来提高,现在我们将考虑腔泵浦强度对系统动力学效应的影响。图 3.2.6 绘制了初始条件相同时腔内光子数$|a|^2$和布居数反转σ_z随归一化时间$\omega_m t$变化的曲线,其中腔泵浦强度分别取不同值$E_p/(2\pi)=1.5\,\mathrm{MHz}$、$2\,\mathrm{MHz}$、$2.5\,\mathrm{MHz}$。可以清楚地看到,$|a|^2$和$\sigma_z$经过一段时间的阻尼振荡后达到稳态解。当$E_p/(2\pi)$从$1.5\,\mathrm{MHz}$、$2\,\mathrm{MHz}$增加到$2.5\,\mathrm{MHz}$时,平均腔内光子数从 0.043、0.084 增加到 0.166,这在图 3.2.3(a)中分别对应于点 C(1.5,0.043)、点 A(2,0.084)和点 D(2.5,0.166)。同时,粒子数反转的稳态解σ_z从-0.950、-0.906增加到-0.831,在图 3.2.3(b)中分别对应于$C'(1.5,-0.950)$、点$A'(2,-0.906)$以及点$D'(2.5,-0.831)$。由方程(3.2.21)和方程(3.2.22)给出的解析结果与微分方程(3.2.11)~方程(3.2.15)给出的数值结果非常一致。

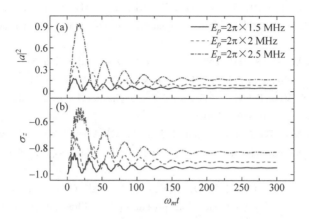

图 3.2.6 $a|_{t=0}=0$ 时(a)腔内光子数 $|a|^2$ 和(b)布居数反转 σ_z 随归一化时间 $\omega_m t$ 变化的曲线

(请扫Ⅱ页二维码看彩图)

3.2.5 小结

本节研究了含有二能级原子的混杂光力系统中平均腔内光子数和原子布居数反转的双稳态行为[39]。原子的加入使得人们能够以更加灵活的方式控制光学双稳态。我们展示了如何通过调节原子驱动场的拉比频率、腔-原子耦合强度以及腔泵浦场的振幅和频率来控制双稳态行为。由于原子引入的强非线性,该系统中可以在腔内光子数较低时出现双稳态行为。此外,研究表明腔场的初始条件对确定腔内光子数和原子布居数反转的稳态解有着重要的影响。我们还简要讨论了腔泵浦场的振幅对系统动力学效应的影响。

参 考 文 献

[1] KIPPENBERG T J, VAHALA K J. Cavity optomechanics: backaction at the mesoscale[J]. Science,2008,321: 1172-1176.
[2] MARQUARDT F,GIRVIN S M. Optomechanics[J]. Physics,2009,2: 40.
[3] ASPELMEYER M, KIPPENBERG T J, MARQUARDT F. Cavity optomechanics [J]. Rev. Mod. Phys. ,2014,86: 1391-1452.
[4] TEUFEL J D,DONNER T,LI D,et al. Sideband cooling of micromechanical motion to the quantum ground state[J]. Nature,2011,475: 359-363.

[5] CHAN J, ALEGRE T P, SAFAVI-NAEINI A H, et al. Laser cooling of a nanomechanical oscillator into its quantum ground state[J]. Nature, 2011, 478: 89-92.

[6] AGARWAL G S, HUANG S. Electromagnetically induced transparency in mechanical effects of light[J]. Phys. Rev. A, 2010, 81: 041803.

[7] WEIS S, RIVIÈRE R, DELÉGLISE S, et al. Optomechanically induced transparency [J]. Science, 2010, 330: 1520-1523.

[8] SAFAVI-NAEINI A H, ALEGRE T P M, CHAN J, et al. Electromagnetically induced transparency and slow light with optomechanics[J]. Nature, 2011, 472: 69-73.

[9] DORSEL A, MCCULLEN J D, MEYSTRE P, et al. Optical bistability and mirror confinement induced by radiation pressure[J]. Phys. Rev. Lett., 1983, 51: 1550-1553.

[10] JIANG C, LIU H X, CUI Y S, et al. Controllable optical bistability based on photons and phonons in a two-mode optomechanical system[J]. Phys. Rev. A, 2013, 88: 055801.

[11] LUDWIG M, SAFAVI-NAEINI A H, PAINTER O, et al. Enhanced quantum nonlinearities in a two-mode optomechanical system[J]. Phys. Rev. Lett., 2012, 109: 063601.

[12] XUEREB A, GENES C, DANTAN A. Strong coupling and long-range collective interactions in optomechanical arrays[J]. Phys. Rev. Lett., 2012, 109: 223601.

[13] KÓMÁR P, BENNETT S D, STANNIGEL K, et al. Single-photon nonlinearities in two-mode optomechanics[J]. Phys. Rev. A, 2013, 87: 013839.

[14] LEMONDE M A, DIDIER N, CLERK A A. Nonlinear interaction effects in a strongly driven optomechanical cavity[J]. Phys. Rev. Lett., 2013, 111: 053602.

[15] LIAO J Q, LAW C K, KUANG L M, et al. Enhancement of mechanical effects of single photons in modulated two-mode optomechanics[J]. Phys. Rev. A, 2015, 92: 013822.

[16] LEMONDE M A, DIDIER N, CLERK A A. Enhanced nonlinear interactions in quantum optomechanics via mechanical amplification[J]. Nat. Commun., 2016, 7: 11338.

[17] RESTREPO J, CIUTI C, FAVERO I. Single-polariton optomechanics[J]. Phys. Rev. Lett., 2014, 112: 013601.

[18] HEIKKILA T T, MASSEL F, TUORILA J, et al. Enhancing optomechanical coupling via the Josephson effect[J]. Phys. Rev. Lett., 2014, 112: 203603.

[19] PIRKKALAINEN J M, CHO S U, LI J, et al. Hybrid circuit cavity quantum electrodynamics with a micromechanical resonator[J]. Nature, 2013, 494: 211-215.

[20] RAMOS T, SUDHIR V, STANNIGEL K, et al. Nonlinear quantum optomechanics via individual intrinsic two-level defects[J]. Phys. Rev. Lett., 2013, 110: 193602.

[21] PFLANZER A C, ROMERO-ISART O, CIRAC J I. Optomechanics assisted by a qubit: from dissipative state preparation to many-partite systems[J]. Phys. Rev. A, 2013, 88: 033804.

[22] PIRKKALAINEN J M, CHO S U, MASSEL F, et al. Cavity optomechanics mediated by a quantum two-level system[J]. Nat. Commun. , 2015, 6: 6981.

[23] LECOCQ F, TEUFEL J D, AUMENTADO J, et al. Resolving the vacuum fluctuations of an optomechanical system using an artificial atom[J]. Nat. Phys. , 2015, 11: 635-639.

[24] WANG H, SUN H C, ZHANG J, et al. Transparency and amplification in a hybrid system of mechanical resonator and circuit QED[J]. Sci. China Phys. Mech. Astron. , 2012, 55: 2264-2272.

[25] AKRAM M J, KHAN M M, SAIF F. Tunable fast and slow light in a hybrid optomechanical system[J]. Phys. Rev. A, 2015, 92: 023846.

[26] SETE E A, ELEUCH H. Controllable nonlinear effects in an optomechanical resonator containing a quantum well[J]. Phys. Rev. A, 2012, 85: 043824.

[27] KANAMOTO R, MEYSTRE P. Optomechanics of a quantumdegenerate Fermi gas [J]. Phys. Rev. Lett. , 2010, 104: 063601.

[28] PURDY T P, BROOKS D W C, BOTTER T, et al. Tunable cavity optomechanics with ultracold atoms[J]. Phys. Rev. Lett. , 2010, 105: 133602.

[29] GUPTA S, MOORE K L, MURCH K W, et al. Cavity nonlinear optics at low photon numbers from collective atomic motion [J]. Phys. Rev. Lett. , 2007, 99: 213601.

[30] ZHANG J M, CUI F C, ZHOU D L, et al. Nonlinear dynamics of a cigar-shaped Bose-Einstein condensate in an optical cavity[J]. Phys. Rev. A, 2009, 79: 033401.

[31] ZHANG K, CHEN W, BHATTACHARYA M, et al. Hamiltonian chaos in a coupled BEC-coptomechanical-cavity system[J]. Phys. Rev. A, 2010, 81: 013802.

[32] YANG S, AMRI M, ZUBAIRY M S. Anomalous switching of optical bistability in a Bose-Einstein condensate[J]. Phys. Rev. A, 2013, 87: 033836.

[33] BRENNECKE F, RITTER S, DONNER T, et al. Cavity optomechanics with a Bose-Einstein condensate[J]. Science, 2008, 322: 235-238.

[34] YASIR K A, LIU W M. Tunable bistability in hybrid Bose-Einstein condensate optomechanics[J]. Sci. Rep. , 2015, 5: 10612.

[35] YAN D, WANG Z H, REN C N, et al. Duality and bistability in an optomechanical cavity coupled to a Rydberg superatom[J]. Phys. Rev. A, 2015, 91: 023813.

[36] XIONG W, JIN D Y, QIU Y Y, et al. Cross-Kerr effect on an optomechanical system[J]. Phys. Rev. A, 2016, 93: 023844.

[37] HAMMERER K, WALLQUIST M, GENES C, et al. Strong coupling of a mechanical oscillator and a single atom[J]. Phys. Rev. Lett. , 2009, 103: 063005.

[38] DEJESUS E X, KAUFMAN C. Routh-Hurwitz criterion in the examination of

eigenvalues of a system of nonlinear ordinary differential equations[J]. Phys. Rev. A,1987,35: 5288-5290.

[39] JIANG C,BIAN X T,CUI Y S,et al. Optical bistability and dynamics in an optomechanical system with a two-level atom[J]. J. Opt. Soc. Am. B,2016,33(10): 2099-2104.

3.3 二能级原子与机械振子耦合的混杂光力系统中的光学双稳态和四波混频

3.3.1 引言

包含纳米机械振子的混杂量子系统在过去十几年中受到了广泛的研究[1]。这些工作主要由于混杂系统可用来研究宏观物体的量子性质,并且在量子信息科学和纳米尺度的传感等方面有着重要的应用。通过将机械振子耦合于其他量子物体已经实现了各种各样的混杂系统,包括耦合于光学微腔[2-4]、超导传输线微波腔[5-7],以及由原子、量子点、量子比特和缺陷等构成的二能级系统[8-14]。通过辐射压实现的力学和光学自由度之间的相互作用是近年来迅速发展的腔光力学系统研究中的主要内容[2-7]。最近,该领域中取得了许多非常重要的进展,包括纳米机械振子的基态冷却[15,16]、光力诱导透明[17-19]、超灵敏的传感等[20,21]。此外,包含一个二能级系统的混杂光力系统被理论上提出[22-28],并在实验上实现[29-31]。

在腔光力学领域,多塞尔(Dorsel)等首先发现可以通过一束强的激光场控制光学双稳态[32]。最近,包含量子阱[33]、超冷原子[34-36]、玻色-爱因斯坦凝聚体(BEC)[37,38]的光力系统中的光学双稳态得到了广泛的研究。熊(Xiong)等[39]和达拉菲(Dalafi)等[40]理论上讨论了交叉克尔效应对稳态光子数的双稳态行为的影响。在我们之前的工作中,研究了一种混杂光力系统中的光学双稳态和动力学效应,其中该混杂系统由腔场同时耦合于一个机械振子和一个二能级系统构成[41]。此外,在一束强的控制场作用下,光系统对一束弱的探测场的光学响应会被改变,导致光力诱导透明[17-19]和双色光力诱导透明现象[23]的出现。光力诱导透明类似于电磁诱导透明[42,43],而电磁诱导透明可用来大大增强四波混频的强度[44,45]。

基于上述研究进展,本节研究另外一种混杂光力系统中的光学双稳态和四波混频现象,该系统由机械振子一方面通过辐射压耦合于腔肠,另一方面通过 J-C 耦合于一个二能级系统(量子比特)[22-24]。我们发现稳态光子数和声子数的双稳行为可通过控制场的拉比频率以及腔-控制场之间的失谐量进行调节。当量子比特-振子之间的耦合强度小于二能级系统的跃迁频率时,二能级系统对稳态声子数的双稳行为影响可忽略不计。但是,二能级系统可导致该混杂系统中出现双色光机械诱导透明,从而使得四波混频的强度在透明窗口附近被大大增强。

3.3.2 模型和理论

我们考虑如图 3.3.1 所示的光力系统。机械振子通过辐射压力耦合于一个单模的腔场,并且通过 J-C 耦合形式耦合一个二能级系统,腔场和二能级系统之间没有直接的相互作用。该混杂系统的哈密顿量可表示为

$$H_0 = \hbar\omega_a a^\dagger a + \hbar\omega_b b^\dagger b + \frac{\hbar}{2}\omega_q \sigma_z - \hbar\chi a^\dagger a(b^\dagger + b) + \hbar g(b^\dagger \sigma_- + b\sigma_+)$$

(3.3.1)

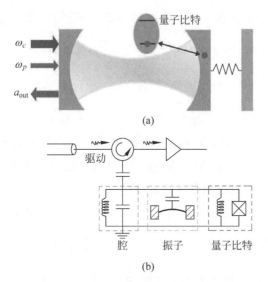

图 3.3.1 混杂光力系统示意图

(a) 光力腔的一个镜子固定而另外一个可在其平衡位置附近振动(看作机械振子),腔场通过辐射压耦合于机械振子。此外,机械振子又耦合于一个二能级系统(量子比特)。一束强的控制场和一束弱的探测场同时驱动腔场;(b) 等效电路

(请扫Ⅱ页二维码看彩图)

式中,ω_a 是腔场的共振频率,$a^\dagger(a)$ 是腔场的产生(湮灭)算符;ω_b 是机械振子的共振频率,$b^\dagger(b)$ 是振子的产生(湮灭)算符;ω_q 是二能级系统基态和激发态之间的跃迁频率,$\sigma_z = |e\rangle\langle e| - |g\rangle\langle g|$ 是泡利算符,$\sigma_+(\sigma_-)$ 是二能级系统的上升(下降)算符;χ 是腔场和机械振子之间的耦合强度,g 是机械振子和二能级系统之间的耦合强度。

为了研究混杂系统的光学响应,我们考虑腔场同时受到一束频率为 ω_c 的强的控制场和一束频率为 ω_p 的弱的探测场的驱动。在控制场频率 ω_c 的旋转框架下,该混杂光力系统的哈密顿量可表示为[23]

$$H = \hbar\Delta_a a^\dagger a + \hbar\Delta_b b^\dagger b + \frac{\hbar}{2}\omega_q \sigma_z - \hbar\chi a^\dagger a(b^\dagger + b) + \hbar g(b^\dagger\sigma_- + b\sigma_+) + i\hbar(\Omega a^\dagger - \Omega^* a) + i\hbar(\varepsilon e^{-i\Delta t}a^\dagger - \varepsilon^* e^{i\Delta t}a) \quad (3.3.2)$$

式中,$\Delta_a = \omega_a - \omega_c$ 是腔场和控制场之间的失谐量,$\Delta = \omega_p - \omega_c$ 是探测场和控制场之间的失谐量,Ω 和 ε 分别表示强的控制场和弱的探测场的拉比频率。

根据海森伯运动方程并引入相应的衰减和噪声项,海森伯-朗之万方程可表示为

$$\dot{a} = -(\gamma_a + i\Delta_a a)a + i\chi a(b^\dagger + b) + \Omega + \varepsilon e^{-i\Delta t} + \sqrt{2\gamma_b}\,a_{in}(t) \quad (3.3.3)$$

$$\dot{b} = -(\gamma_b + i\omega_b)b + i\chi a^\dagger a - ig\sigma_- + \sqrt{2\gamma_b}\,b_{in}(t) \quad (3.3.4)$$

$$\dot{\sigma}_- = -\left(\frac{\gamma_q}{2} + i\omega_q\right)\sigma_- - igb\sigma_z + \sqrt{\gamma_q}\,\Gamma_-(t) \quad (3.3.5)$$

$$\dot{\sigma}_z = -\gamma_q(\sigma_z + 1) - 2ig(b\sigma_+ - b^\dagger\sigma_-) + \sqrt{\gamma_q}\,\Gamma_z(t) \quad (3.3.6)$$

式中,γ_a、γ_b 和 γ_q 分别表示腔场、机械振子和二能级系统的衰减率。算符 $a_{in}(t)$、$b_{in}(t)$ 和 $\Gamma(t)$ 表示相应的环境噪声,它们的平均值为零。利用平均场近似,我们可以得到以下的期待值方程:

$$\langle\dot{a}\rangle = -(\gamma_a + i\Delta_a a)\langle a\rangle + i\chi\langle a\rangle(\langle b^\dagger\rangle + \langle b\rangle) + \Omega + \varepsilon e^{-i\Delta t} \quad (3.3.7)$$

$$\langle\dot{b}\rangle = -(\gamma_b + i\omega_b)\langle b\rangle + i\chi\langle a^\dagger\rangle\langle a\rangle - ig\langle\sigma_-\rangle \quad (3.3.8)$$

$$\langle\dot{\sigma}_-\rangle = -\left(\frac{\gamma_q}{2} + i\omega_q\right)\langle\sigma_-\rangle - ig\langle b\rangle\langle\sigma_z\rangle \quad (3.3.9)$$

$$\langle\dot{\sigma}_z\rangle = -\gamma_q(\langle\sigma_z\rangle + 1) - 2ig(\langle b\rangle\langle\sigma_+\rangle - \langle b^\dagger\rangle\langle\sigma_-\rangle) \quad (3.3.10)$$

为了求解非线性方程(3.3.7)~方程(3.3.10),使用以下的代换[46]:

$$\langle a(t)\rangle = A_0 + A_+ e^{i\Delta t} + A_- e^{-i\Delta t} \quad (3.3.11)$$

$$\langle b(t)\rangle = B_0 + B_+ e^{i\Delta t} + B_- e^{-i\Delta t} \tag{3.3.12}$$

$$\langle \sigma_-(t)\rangle = L_0 + L_+ e^{i\Delta t} + L_- e^{-i\Delta t} \tag{3.3.13}$$

$$\langle \sigma_z(t)\rangle = Z_0 + Z_+ e^{i\Delta t} + Z_- e^{-i\Delta t} \tag{3.3.14}$$

式中,A_0、B_0、L_0、Z_0 分别是 a、b、σ_-、σ_z 在没有探测场时的稳态解。对于弱的探测场,$A_\pm$、$B_\pm$、$L_\pm$、$Z_\pm$ 远小于相应的稳态解。将方程(3.3.11)~方程(3.3.14)代入方程(3.3.7)~方程(3.3.10),可以得到以下稳态解:

$$A_0 = \frac{\Omega}{\gamma_a + i\Delta_a - i\chi(B_0 + B_0^*)}, \quad B_0 = \frac{\chi|A_0|^2 - gL_0}{\omega_b - i\gamma_b}$$

$$L_0 = \frac{2gB_0 Z_0}{2\omega_q - i\gamma_q}, \quad Z_0 = -\frac{\gamma_q^2 + 4\omega_q^2}{\gamma_q^2 + 4\omega_q^2 + 8g^2|B_0|^2} \tag{3.3.15}$$

$$A_- = \frac{\alpha_9 \varepsilon}{\alpha_8 \alpha_9 - \alpha_{10}}, \quad A_+ = \frac{i\chi(\alpha_6 + \alpha_7^*)A_0^2 \varepsilon^*}{\alpha_8^* \alpha_9^* - \alpha_{10}^*} \tag{3.3.16}$$

式中,

$$\alpha_1 = 2g/(\Delta - i\gamma_q), \quad \alpha_2 = \frac{1}{\beta_3}[igZ_0\beta_1 - g\alpha_1 B_0 L_0^*(i\beta_1 - g\alpha_1|B_0|^2)]$$

$$\alpha_3 = i\frac{g\alpha_1 B_0}{\beta_3}(ig\alpha_1|B_0|^2 L_0 + igB_0 Z_0 + \beta_1 L_0)$$

$$\alpha_4 = \frac{1}{\beta_3^*}[igZ_0\beta_2 + g\alpha_1^* B_0 L_0^*(i\beta_2 + g\alpha_1^*|B_0|^2)]$$

$$\alpha_5 = i\frac{g\alpha_1^* B_0}{\beta_3^*}(ig\alpha_1^*|B_0|^2 L_0 - igB_0 Z_0 - \beta_2 L_0)$$

$$\alpha_6 = \frac{-g\chi\alpha_3 + i\chi\beta_4}{\beta_4 \beta_5 - g^2\alpha_3 \alpha_5^*}, \quad \alpha_7 = \frac{-g\chi\alpha_5 + i\chi\beta_5^*}{\beta_4^* \beta_5^* - g^2\alpha_3^* \alpha_5} \tag{3.3.17}$$

$$\alpha_8 = \gamma_a + i[\Delta_a - \Delta - \chi(B_0 + B_0^*) - \chi(\alpha_6^* + \alpha_7)|A_0|^2]$$

$$\alpha_9 = \gamma_a - i[\Delta_a + \Delta - \chi(B_0 + B_0^*) - \chi(\alpha_6^* + \alpha_7)|A_0|^2]$$

$$\alpha_{10} = \chi^2(\alpha_6^* + \alpha_7)^2|A_0|^4$$

$$\beta_1 = \gamma_q/2 - i\omega_q - ig\alpha_1|B_0|^2 + i\Delta$$

$$\beta_2 = \gamma_q/2 - i\omega_q + ig\alpha_1^*|B_0|^2 - i\Delta$$

$$\beta_3 = (\gamma_q/2 + i\Delta)^2 - 2ig\alpha_1|B_0|^2(\gamma_q/2 + i\Delta)^2 + \omega_q^2$$

$$\beta_4 = \gamma_b - i(\omega_b - \Delta + g\alpha_4^*), \quad \beta_5 = \gamma_b + i(\omega_b + \Delta + g\alpha_2)$$

根据方程(3.3.15),我们可以得到腔场的稳态光子数$|A_0|^2$和机械振子的稳态声子数$|B_0|^2$满足以下方程:

$$|A_0|^2\left\{\gamma_a^2+\left[\Delta_a-\frac{2\chi^2|A_0|^2(2\omega_q\nu_1+\gamma_q\nu_2)}{\nu_1^2+\nu_2^2}\right]^2\right\}=\Omega^2 \quad (3.3.18)$$

$$|B_0|^2(\nu_1^2+\nu_2^2)=\chi^2|A_0|^4[(2\omega_q\nu_1+\gamma_q\nu_2)^2+(\gamma_q\nu_1-2\omega_1\nu_1)^2] \quad (3.3.19)$$

式中,$\nu_1=2\omega_b\omega_q-\gamma_b\gamma_q+2g^2Z_0$,$\nu_2=\omega_b\gamma_q+2\gamma_b\omega_q$,$Z_0=-\dfrac{\gamma_q^2+4\omega_q^2}{\gamma_q^2+4\omega_q^2+8g^2|B_0|^2}$。因此,稳态时腔内光子数$|A_0|^2$、力学模式的声子数$|B_0|^2$以及二能级系统的布居数反转$Z_0$三者之间互相依赖。这种形式的耦合方程具有光学多稳态的特征[34,36]。

3.3.3 光子数和声子数的双稳行为

本节我们首先根据方程(3.3.18)和方程(3.3.19)研究稳态时腔内光子数和机械振子的声子数的双稳行为。为了观察到双稳行为,我们可以得到$\partial\Omega^2/\partial|A_0|^2=0$[33],从而

$$\gamma_a^2+\Delta_a^2-8\Delta_a\eta\chi^2|A_0|^2+12\eta^2\chi^4|A_0|^4=0 \quad (3.3.20)$$

式中,$\eta=\dfrac{2\omega_q\nu_1+\gamma_q\nu_2}{\nu_1^2+\nu_2^2}$。双稳条件可在方程(3.3.20)含有两个不同的正解时得到。但是,参数η包含ν_1,这又与Z_0、$|A_0|^2$、$|B|^2$相关。因此,双稳条件的解析表达式非常复杂,这里就不再给出。接下来将展示我们的数值结果,我们采用与文献[22]中相似的参数:$\omega_b/(2\pi)=\omega_q/(2\pi)=5\text{ GHz}$,$\gamma_b=6\times10^{-6}\omega_b$,$\gamma_a=0.05\omega_b$,$\gamma_q/(2\pi)=1\text{ MHz}$。量子比特与机械振子之间的耦合强度$g$可达$g=0.1\omega_b$[12,29],光力耦合强度认为与量子比特-振子之间的耦合强度相等[22]。

图3.3.2(a)所示是不同腔-控制场之间失谐量Δ_a时稳态光子数$|A_0|^2$随控制场的归一化拉比频率Ω/ω_b的变化曲线。对于固定的失谐量Δ_a,可以看出当拉比频率在一定范围内变化时,稳态光子数最多可以有三个不同的数值。在这种情况下,方程(3.3.18)关于$|A_0|^2$有三个实根,其中最大和最小的根是稳定的而中间的根是不稳定的。因此,在控制场拉比频率变化时我们可以看到光子数的双稳行为。此外,当Δ_a从$0.6\omega_b$增加到$0.8\omega_b$和ω_b时,观

察到双稳行为时的拉比频率的阈值变大,并且双稳区域变得更宽。图 3.3.2(b)所示是稳态光子数随腔控制场失谐量变化时的情况。可以看出,当失谐量从 -1 增加到临界值 C_1 时,光子数单调递增。但是,当失谐量在一定范围内时,双稳行为出现,其中上面的一支和下面的一支是稳定的。当我们渐渐增加失谐量时,稳态光子数开始沿着曲线的上面一支变化。当失谐量达到另外一个临界点 C_2 时,光子数将跳跃到下面的一支。如果再继续减小失谐量,光子数将沿着下面一支变化;直到失谐量又减小为临界点 C_1 时光子数又会跳回上面一支。

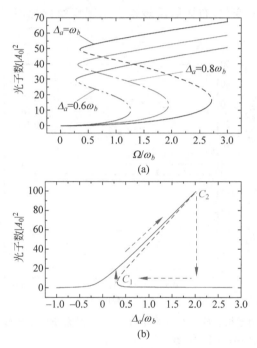

图 3.3.2 稳态腔内光子数随着(a)归一化拉比频率和(b)归一化失谐量变化的曲线

(请扫Ⅱ页二维码看彩图)

此外,我们可以从方程(3.3.19)看出力学模式的稳态声子数 $|B_0|^2$ 和二能级系统的布居数反转 Z_0 是与腔内光子数 $|A_0|^2$ 相关的。图 3.3.3(a)和(b)所示分别是稳态声子数和布居数反转随归一化拉比频率变化的情况,其中量子比特-振子耦合强度分别取为 0、$0.1\omega_b$ 和 $0.5\omega_b$。对于确定的腔控制场失谐量,当控制场的拉比频率变化时,稳态声子数也会出现双稳行为,变化

趋势跟图 3.3.2(a) 中 $\Delta_a = \omega_b$ 的曲线类似。此外,可以看出图 3.3.3(a) 中 $g=0$ 和 $g=0.1\omega_b$ 这两条曲线基本重合。而当 $g=0.5\omega_b$ 时曲线中的双稳区域也只是有稍微一点不同。因此,当耦合强度 g 比机械振子的共振频率小时,二能级系统对声子数双稳行为的影响可以忽略不计。但是,如果耦合强度 g 变得大于振子共振频率时,方程(3.3.18)和方程(3.3.19)最多可以出现五个实根,因此将出现多稳行为。此外,我们在图 3.3.3(b) 中研究耦合强度 g 对布居数反转 Z_0 的影响。当量子比特和机械振子之间没有相互作用时,二能级系统停留在它的基态上。但是当量子比特-振子之间的耦合打开时,随着控制场拉比频率的变化,布居数反转也会出现双稳行为,并且耦合强度 g 的影响非常明显。

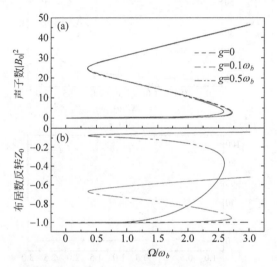

图 3.3.3 (a) 力学模式的稳态声子数和 (b) 二能级系统的布居数反转随归一化拉比频率变化的曲线

(请扫Ⅱ页二维码看彩图)

3.3.4 共振增强的四波混频过程

为了研究输出场的光学性质,我们利用输入-输出理论 $\langle a_{\text{out}} \rangle + \Omega + \varepsilon e^{-i\Delta t} = \gamma_{a,e} \langle a \rangle$ [47] 得到输出场的期待值,表达式如下:

$$\langle \tilde{a}_{\text{out}} \rangle = (\gamma_{a,e} A_0 - \Omega) e^{-i\omega_c t} + (\gamma_{a,e} A_- - \varepsilon) e^{-i(\omega_c - \Delta)t} + \gamma_{a,e} A_+ e^{-i(\omega_c - \Delta)t}$$

$$=(\gamma_{a,e}A_0-\Omega)\mathrm{e}^{-\mathrm{i}\omega_c t}+\underbrace{(\gamma_{a,e}A_--\varepsilon)\mathrm{e}^{-\mathrm{i}\omega_p t}}_{\text{探测场}}+\underbrace{\gamma_{a,e}A_+\mathrm{e}^{-\mathrm{i}(2\omega_c-\omega_p)t}}_{\text{四波混频场}}$$

$$\underbrace{\phantom{(\gamma_{a,e}A_0-\Omega)\mathrm{e}^{-\mathrm{i}\omega_c t}}}_{\text{控制场}}$$

(3.3.21)

式中，$\tilde{a}_{\text{out}}$ 是在原始坐标下的输出场算符，$\gamma_{a,e}$ 是腔场由于外部耦合引起的衰减率。从方程(3.3.21)可以看出，输出场中除了包含控制场和探测场这两个分量外，还包含一个新产生的频率为 $2\omega_c-\omega_p$ 的四波混频分量。探测场透射谱定义为

$$t_p=\frac{\gamma_{a,e}A_--\varepsilon}{\varepsilon}=\frac{\gamma_{a,e}\alpha_9}{\alpha_8\alpha_9-\alpha_{10}}-1 \quad (3.3.22)$$

此外，以探测场为单位的归一化的四波混频强度定义为

$$\mathrm{FWM}=\left|\frac{\gamma_{a,e}A_+}{\varepsilon^*}\right|=\left|\frac{\mathrm{i}\chi\gamma_{a,e}(\alpha_6+\alpha_7^*)A_0^2}{\alpha_8^*\alpha_9^*-\alpha_{10}^*}\right| \quad (3.3.23)$$

图 3.3.4(a)和(b)所示分别是探测场透射率 $|t_p|^2$ 和四波混频强度随探测-控制场失谐量 Δ（以机械振子共振频率为单位）变化的情况，其中 $\Delta_a=\omega_b$，$\Omega=0.1\omega_b$。在这种弱激发情况下，系统不会出现双稳行为，并且稳态光子数和声子数都小于 1，如图 3.3.2(a)和图 3.3.3(a)所示。在没有二能级作用时，从图 3.3.4(a)可以看出，在探测场透射谱的中间出现了光力诱导透明现象，而这一现象可以通过振荡频率为探测场和控制场的拍频的辐射压来进行很好的解释。此外，图 3.3.4(b)表明，在光力诱导透明的条件下，四波混频的强

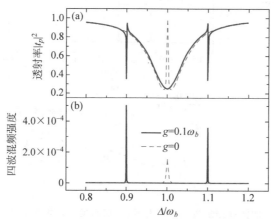

图 3.3.4 (a)探测场透射谱和(b)四波混频强度随探测-控制失谐量变化的曲线

（请扫Ⅱ页二维码看彩图）

度可以被大大增强。四波混频谱在 $\Delta/\omega_b=1$ 处有一个尖峰。当量子比特-振子的耦合强度打开时,在 $\Delta/\omega_b=1$ 处的透明窗口分裂为两个对称的,即出现了双色光力诱导透明现象。相应地,图 3.3.4(b) 中在透明窗口的位置处出现了两个尖峰。四波混频强度的峰值相比于没有二能级系统时的情况也大大增强了。

为了能确定四波混频谱中峰值的位置,在图 3.3.5(a) 中画了不同量子比特-振子耦合强度时四波混频强度随归一化探测-控制场失谐量变化的情况。我们发现两个峰值之间的距离随着振子和二能级系统之间耦合强度的增加而变大。图 3.3.5(a) 的插图表明峰劈裂距离与耦合强度 g 之间有着线性关系,从而为根据四波混频谱中两个峰值之间的距离探测耦合强度 g 提供了一种方法。我们也在数值上验证了在现有参数条件下这样的线性关系在归一化拉比频率 $\Omega/\omega_b \leqslant 0.3$ 时仍然有效。此外,从图 3.3.5(a) 中可以看出这两个峰分别位于 $\Delta=\omega_b+g$ 和 $\Delta=\omega_b-g$。需要指出的是,图 3.3.5(a) 中峰的线宽非常小是因为机械振子的衰减率非常小。一种解释峰值位置的直观的物理图像可以由图 3.3.5(b) 所示的能级图给出。没有二能级系统时,$|0_a,0_b\rangle$、

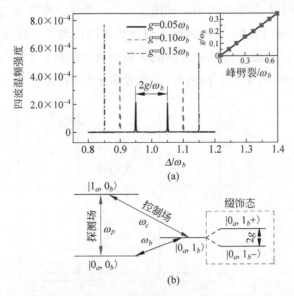

图 3.3.5 (a) 四波混频强度随归一化探测-控制场失谐量变化的曲线;(b) 腔被驱动至红边带时该混杂系统的能级图

(请扫Ⅱ页二维码看彩图)

$|0_a,1_b\rangle$、$|1_a,0_b\rangle$ 三个态构成了 Λ 型三能级结构，其中下标 a 和 b 分别表示光子和声子态。但是，当一个跃迁频率 ω_q 等于机械振子共振频率 ω_b 的二能级系统耦合于机械振子时，态 $|0_a,1_b\rangle$ 分裂为频率为 ω_b+g 的态 $|0_a,1_b+\rangle$ 和频率为 ω_b-g 的态 $|0_a,1_b-\rangle$[48]。这里 $|1_b\pm\rangle$ 代表单声子态和二能级系统形成的缀饰态，当 $\omega_q=\omega_b$ 时可表示为 $|1_b\pm\rangle=(|1_b,g\rangle\pm|0_b,e\rangle)/\sqrt{2}$。当腔场被一束强的控制场驱动至红边带，即 $\Delta_a=\omega_a-\omega_c=\omega_b$ 时，振动频率为拍频 Δ 的辐射压将产生由于修饰态形成的斯托克斯和反斯托克斯场。反斯托克斯场的频率为 $\omega_{as,1}=\omega_c+(\omega_b+g)=\omega_a+g$ 和 $\omega_{as,2}=\omega_c+(\omega_b-g)=\omega_a-g$，正好和腔场频率近共振，但是斯托克斯场由于和腔场高度非共振而被强烈抑制。近简并的探测场和产生的反斯托克斯场之间的相消干涉导致了图 3.3.4(a) 的探测场透射谱中 $\Delta=\omega_b+g$ 和 $\Delta=\omega_b-g$ 两个透明窗口的出现，从而又进一步使得这两个位置的四波混频强度被大大增强。但是，当控制场的拉比频率变得很强以致多声子激发出现时，四波混频谱中两个峰值的位置将发生移动，并且这两个峰之间的距离将变得更大。

接下来，我们将讨论如何控制四波混频的强度。在光机械诱导透明出现的条件下，我们画出了 $\Delta=\omega_b-g$ 处的四波混频强度随控制场归一化的拉比频率变化的情况，如图 3.3.6 所示。可以看出，当拉比频率变大时四波混频的峰值强度开始增加很快，然后达到一个最大值。进一步增加拉比频率时，四波混频强度开始变小并最终达到一个几乎稳定的值。产生这一结果的物理原因跟混杂系统的非线性有关。当控制场的拉比频率相对较小时，增加拉比

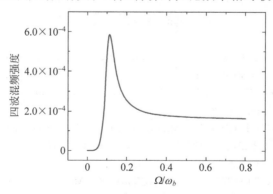

图 3.3.6 四波混频的强度随着归一化拉比频率变化的曲线

(请扫 Ⅱ 页二维码看彩图)

频率将使得腔内的光子数增加,从而使得腔场与机械振子之间的有效耦合强度变大。因此,探测场和控制场之间可以通过修饰后的机械振子产生更高效混杂形成四波混频分量。但是,当拉比频率进一步增加时,光学双稳态将出现,并且多声子激发出现时,四波混频谱中两个峰值的位置将发生一定的移动使得峰之间的距离变大,这些都会使该混杂系统中的四波混频过程变得更加复杂。

3.3.5 小结

我们研究了一种混杂光力系统中的光学双稳态和四波混频过程[49],该系统由一个机械振子同时耦合于一个腔场和一个二能级系统构成。数值结果表明,光学双稳态可通过驱动腔的控制场进行有效调控。如果二能级-振子之间的耦合强度比二能级的跃迁频率小时,耦合强度 g 对稳态光子数和声子数的双稳行为的影响可以忽略,但 g 将会对二能级的布居数反转的双稳行为产生显著的影响。此外,二能级系统的出现明显改变了腔的输出场的光学性质。探测场的透射谱中出现了两个透明窗口,从而进一步增加了四波混频的强度。四波混频谱中两个峰之间的距离可以通过二能级-振子之间的耦合强度进行调制。本节最后还研究了四波混频强度的峰值和控制场的拉比频率之间的关系。

参 考 文 献

[1] POOT M, VAN DER ZANT H S J. Mechanical systems in the quantum regime[J]. Phys. Rep. ,2012,511:273.

[2] KIPPENBERG T J, VAHALA K J. Cavity optomechanics:back-action at the mesoscale[J]. Science,2008,321:1172.

[3] MARQUARDT F,GIRVIN S M. Optomechanics[J]. Physics,2009,2:40.

[4] ASPELMEYER M, KIPPENBERG T J, MARQUARDT F. Cavity optomechanics[J]. Rev. Mod. Phys. ,2014,86:1391.

[5] REGAL C A, TEUFEL J D, LEHNERT K W. Measuring nanomechanical motion with a microwave cavity interferometer[J]. Nat. Phys. ,2008,4:555-560.

[6] ROCHELEAU T,NDUKUM T,MACKLIN C,et al. Preparation and detection of a mechanical resonator near the ground state of motion[J]. Nature,2010,463:72-75.

[7] TEUFEL J D, LI D, ALLMAN M S, et al. Circuit cavity electromechanics in the strong-coupling regime[J]. Nature,2011,471: 204-208.

[8] HAMMERER K,WALLQUIST M,GENES C,et al. Strong coupling of a mechanical oscillator and a single atom[J]. Phys. Rev. Lett. ,2009,103: 063005.

[9] YEO I,DE ASSIS P L,GLOPPE A,et al. Strain-mediated coupling in a quantum dot-mechanical oscillator hybrid system[J]. Nat. Nanotechnol. ,2014,9: 106-110.

[10] WILSON-RAE I,ZOLLER P,IMAMOGLU A. Laser cooling of a nanomechanical resonator mode to its quantum ground state [J]. Phys. Rev. Lett. , 2004, 92: 075507.

[11] LAHAYE M D, SUH J, ECHTERNACH P M, et al. Nanomechanical measurements of a superconducting qubit[J]. Nature,2009,459: 960-964.

[12] O'CONNELL A D,HOFHEINZ M,ANSMANN M,et al. Quantum ground state and single-phonon control of a mechanical resonator [J]. Nature, 2010, 464: 697-703.

[13] KOLKOWITZ S,JAYICH A C B,UNTERREITHMEIER Q P,et al. Coherent sensing of a mechanical resonator with a single-spin qubit[J]. Science,2012,335: 1603-1606.

[14] LEE K W, LEE D, OVARTCHAIYAPONG P, et al. Strain coupling of a mechanical resonator to a single quantum emitter in diamond[J]. Phys. Rev. Appl. ,2016,6: 034005.

[15] CHAN J, ALEGRE T P, SAFAVI-NAEINI A H, et al. Laser cooling of a nanomechanical oscillator into its quantum ground state[J]. Nature, 2011, 478: 89-92.

[16] TEUFEL J D,DONNER T,LI D,et al. Sideband cooling of micromechanical motion to the quantum ground state[J]. Nature,2011,475: 359-363.

[17] AGARWAL G S, HUANG S. Electromagnetically induced transparency in mechanical effects of light[J]. Phys. Rev. A,2010,81: 041803.

[18] WEIS S, RIVIERE R , DELEGLISE S, et al. Optomechanically induced transparency[J]. Science,2010,330: 1520-1523.

[19] SAFAVI-NAEINI A H, ALEGRE T P M, CHAN J, et al. Electromagnetically induced transparency and slow light with optomechanics[J]. Nature,2011,472: 69-73.

[20] GAVARTIN E,VERLOT P,KIPPENBERG T J. A hybrid on-chip optomechanical transducer for ultrasensitive force measurements[J]. Nat. Nanotechnol. ,2012,7: 509-514.

[21] JIANG C, CUI Y, ZHU K D. Ultrasensitive nanomechanical mass sensor using hybrid opto-electromechanical systems[J]. Opt. Express,2014,22: 13773-13783.

[22] RAMOS T,SUDHIR V,STANNIGEL K,et al. Nonlinear quantum optomechanics via individual intrinsic two-level defects[J]. Phys. Rev. Lett. ,2013,110: 193602.

[23] WANG H, GU X, LIU Y X, et al. Optomechanical analog of two-color electromagnetically induced transparency: photon transmission through an optomechanical device with a two-level system[J]. Phys. Rev. A, 2014, 90: 023817.

[24] CERNOTIK O, HAMMERER K. Measurement-induced long-distance entanglement of superconducting qubits using optomechanical transducers[J]. Phys. Rev. A, 2016, 94: 012340.

[25] RESTREPO J, CIUTI C, FAVERO I. Single-polariton optomechanics[J]. Phys. Rev. Lett., 2014, 112: 013601.

[26] AKRAM M J, KHAN M M, SAIF F. Tunable fast and slow light in a hybrid optomechanical system[J]. Phys. Rev. A, 2015, 92: 023846.

[27] RESTREPO J, FAVERO I, CIUTI C. Fully coupled hybrid cavity optomechanics: quantum interferences and correlations[J]. Phys. Rev. A, 2017, 95: 023832.

[28] COTRUFO M, FIORE A, VERHAGEN E. Coherent atom-phonon interaction through mode field coupling in hybrid optomechanical systems[J]. Phys. Rev. Lett., 2017, 118: 133603.

[29] PIRKKALAINEN J M, CHO S U, LI J, et al. Hybrid circuit cavity quantum electrodynamics with a micromechanical resonator[J]. Nature, 2013, 494: 211.

[30] PIRKKALAINEN J M, CHO S U, MASSEL F, et al. Cavity optomechanics mediated by a quantum two-level system[J]. Nat. Commun., 2015, 6: 6981.

[31] LECOCQ F, TEUFEL J D, AUMENTADO J, et al. Resolving the vacuum fluctuations of an optomechanical system using an artificial atom[J]. Nat. Phys., 2015, 11: 635.

[32] DORSEL A, MCCULLEN J D, MEYSTRE P, et al. Optical bistability and mirror confinement induced by radiation pressure[J]. Phys. Rev. Lett., 1983, 51: 1550.

[33] SETE E A, ELEUCH H. Controllable nonlinear effects in an optomechanical resonator containing a quantum well[J]. Phys. Rev. A, 2012, 85: 043824.

[34] KANAMOTO R, MEYSTRE P. Optomechanics of a quantum-degenerate Fermi gas[J]. Phys. Rev. Lett., 2010, 104: 063601.

[35] PURDY T P, BROOKS D W C, BOTTER T, et al. Tunable cavity optomechanics with ultracold atoms[J]. Phys. Rev. Lett., 2010, 105: 133602.

[36] GUPTA S, MOORE K L, MURCH K W, et al. Cavity nonlinear optics at low photon numbers from collective atomic motion[J]. Phys. Rev. Lett., 2007, 99: 213601.

[37] ZHANG K, CHEN W, BHATTACHARYA M, et al. Hamiltonian chaos in a coupled BEC-optomechanical-cavity system[J]. Phys. Rev. A, 2010, 81: 013802.

[38] YASIR K A, LIU W M. Tunable bistability in hybrid Bose-Einstein condensate optomechanics[J]. Sci. Rep., 2015, 5: 10612.

[39] XIONG W, JIN D Y, QIU Y Y, et al. Cross-Kerr effect on an optomechanical

system[J]. Phys. Rev. A,2016,93: 023844.

[40] DALAFI A, NADERI M H. Intrinsic cross-Kerr nonlinearity in an optical cavity containing an interacting Bose-Einstein condensate[J]. Phys. Rev. A, 2017, 95: 043601.

[41] JIANG C, BIAN X T, CUI Y S, et al. Optical bistability and dynamics in an optomechanical system with a two-level atom[J]. J. Opt. Soc. Am. B, 2016, 33: 2099-2104.

[42] FLEISCHHAUER M, IMAMOGLU A, MARANGOS J P. Electromagnetically induced transparency: optics in coherent media[J]. Rev. Mod. Phys., 2005, 77: 633-673.

[43] WU Y, YANG X X. Electromagnetically induced transparency in V-, Λ- and cascade-type schemes beyond steady-state analysis[J]. Phys. Rev. A, 2005, 71: 053806.

[44] LI Y Q, XIAO M. Enhancement of nondegenerate four-wave mixing based on electromagnetically induced transparency in rubidium atoms[J]. Opt. Lett., 1996, 21: 1064-1066.

[45] WU Y, SALDANA J, ZHU Y F. Large enhancement of four-wave mixing by suppression of photon absorption from electromagnetically induced transparency[J]. Phys. Rev. A, 2003, 67: 013811.

[46] BOYD R W. Nonlinear optics[M]. New York: Academic, 2010.

[47] WALLS D F, MILBURN G J. Quantum optics[M]. Berlin: Springer, 1994.

[48] ORSZAG M. Quantum optics[M]. Berlin: Springer, 2000.

[49] JIANG L, YUAN X R, CUI Y S, et al. Optical bistability and four-wave mixing in a hybrid optomechanical system[J]. Phys. Lett. A, 2017, 381: 3289-3294.

第 4 章　腔光力学系统中的光力诱导吸收和放大

4.1　机械驱动下多模光力系统中相位依赖的光力诱导吸收

4.1.1　引言

迅速发展的腔光力学领域主要研究光学和力学自由度之间通过辐射压力产生的相互作用。该领域不但为观察宏观物体的量子力学行为提供了一个有效的平台，而且在超灵敏测量和量子信息处理中有着重要的应用[1-3]。典型的腔光力学系统由法布里-珀罗腔构成，其中一个镜子固定不同，另一个镜子可在其平衡位置附近来回振动，从而被看成一个机械振子。在一束强的控制场和一束弱的探测场共同作用下，光力系统的电磁响应将会因为辐射压诱导的机械运动而改变，进而引起光力诱导透明[4-11]、光力诱导吸收[12-14]、参量放大[15]等现象。光力诱导透明和光力诱导吸收分别是光力系统中的类电磁诱导透明[16,17]和电磁诱导吸收[18]，电磁诱导透明和电磁诱导吸收首先在原子气体后来在固体系统中被观察到。类似于电磁诱导透明，光力诱导透明可用于控制光信号的群速度延迟时间[19,20]，以及将光信息存储在寿命比较长的机械振荡中[21,22]。

此外，当机械振子受到附加的相干场驱动时，光力系统中将出现更加复杂的干涉效应。光力系统中施加机械驱动的一个特别的优点是它可以直接产生机械相干，这与在相位相干原子介质系统中通过在微波频率直接驱动产生原子相干类似[23]。同时，机械振子的相干振动还可由振动频率为强控制场和弱探测场之间拍频的辐射压力引起。因此，探测场可以与由机械驱动引起机械振子振动从而导致控制场被散射的分量以及由辐射压诱导的机械振动导致控制场被散射的分量之间相干涉。理论上已有研究表明机械驱动场的相位和振幅可以用来控制探测场透射谱[24,25]、输出探测场的群速度延迟[26]以及二阶边带产生过程[25,27]。实验上,通过在多模光力系统中同时施加光学驱动和机械驱动,人们观察到了级联光学透明以及延长的光学延迟和光学超前[28]。

最近,由不止两个自由度构成的多模光力系统引起了人们的研究兴趣[13,28-31]。一方面,通过将两个不同频率的电磁腔耦合于一个共同的机械振子可以实现不同光场之间的频率转化[32-36]。另一方面,单个的电磁腔可以耦合于多个机械振子,导致两个力学模式之间的杂化[37]、双模压缩态制备[38]以及双光力诱导透明现象[39,40],并且双光力诱导透明现象在包含二能级原子的光力系统中也得到了研究[41]。此外,布克曼(Buchmann)和斯坦伯-库恩(Stamper-Kurn)推导出了由两个非简并的力学模式弱耦合于一个共同的腔场形成的光力系统中的主方程的形式[42]。

我们理论上研究了多模光力系统在一束强的控制场、一束弱的探测场和弱的相干机械驱动场共同作用下的相位依赖的光力诱导吸收现象。我们的研究表明,当 N 个频率稍有不同的机械振子耦合于一个共同的腔场时,探测场透射谱中最多可以出现 N 个吸收凹陷。此外,详细的分析表明,探测场透射谱可以通过腔控制场和机械驱动场进行有效调控。与之前只考虑红边带的包含机械驱动的普通光力系统的研究相比[24-27],本节主要研究一般的多模光力系统在蓝边带驱动时出现的光力诱导吸收和放大现象。

4.1.2 模型和理论

考虑的光力系统由 N 个频率稍有不同的机械振子分别耦合于一个共同的腔场形成,如图 4.1.1 所示。在微波频域(图 4.1.1(a)),微波腔可以由等效电感 L 和等效电容 C 描述,腔的共振频率可表示为 $\omega_0 = 1/\sqrt{LC}$[7]。每个机械振子的运动会引起随时间变化的电容 C_k,从而各自独立地调制腔场的总电容和共振频率。腔场和机械振子之间的相互作用哈密顿量为 $H_{\text{int}} = \sum_{k=1}^{N} \hbar G_k \hat{x}_k \hat{a}^+ \hat{a}$,其中,$G_k = (\omega_0/2C) \partial C_k / \partial x_k$ 表示腔场共振频率随机械振子位移的变化,$\hat{a}(\hat{a}^+)$ 是腔模的湮灭(产生)算符[15,20,30]。在光学频域(图 4.1.1(b)),

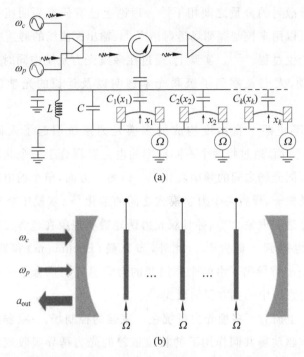

图 4.1.1 由 N 个机械振子分别耦合于一个共同的腔场形成的多模光力系统示意图。腔同时受到强的控制场和弱的探测场驱动,并且每个机械振子均受到频率为控制场和探测场之间拍频的机械驱动

(a) 电路光力系统,其中微波腔由等效电感 L 和等效电容 C 表示。每个机械振子的位移引起随时间变化的电容 C_k 从而改变总电容 C 及腔的频率;(b) 腔内含有 N 个薄膜的光力系统

(请扫Ⅱ页二维码看彩图)

多模光力系统可由包含 N 个薄膜的法布里-珀罗腔构成。在微波和光学频域,我们假设腔场同时受到频率为 ω_c 的强的控制场和频率为 ω_p 的探测场驱动。此外,每个机械振子都受到振幅为 ε_m、频率为 $\Omega = \omega_p - \omega_c$、相位为 ϕ_m 的弱的相干机械泵浦场驱动。在控制场频率 ω_c 的旋转框架下,多模光力系统的哈密顿量可表示为[29,30]

$$H = \hbar \Delta_0 \hat{a}^+ \hat{a} + \sum_{k=1}^{N} \left(\frac{\hat{p}_k^2}{2m_k} + \frac{1}{2} m_k \omega_k^2 \hat{x}_k^2 \right) + \sum_{k=1}^{N} \hbar G_k \hat{x}_k \hat{a}^+ \hat{a} + H_{dr}$$

(4.1.1)

式中,$\Delta_0 = \omega_0 - \omega_c$ 是控制场与腔场频率之间的失谐量,$\hat{x}_k$ 和 $\hat{p}_k$ 分别是第 k 个有效质量为 m_k、频率为 ω_k 机械振子的位置和动量算符,H_{dr} 表示的是光力

系统与驱动场之间的相互作用哈密顿量：

$$H_{dr} = i\hbar\sqrt{\eta_c\kappa/2}\left[(\varepsilon_c + \varepsilon_p e^{-i\Omega t - i\phi_p})\hat{a}^+ - \text{H.c.}\right] - 2\sum_{k=1}^{N}\hat{x}_k\varepsilon_m\cos(\Omega t + \phi_m)$$

(4.1.2)

式中，κ 是腔场总的衰减率，包括内在衰减率 κ_0 和外部衰减率 κ_{ex} 两部分。耦合参数 $\eta_c = \kappa_{ex}/(\kappa_{ex} + \kappa_0)$ 实验上可以调节，本书中我们选取 $\eta_c = 1/2$。ε_c 和 ε_p 分别是控制场和探测场的强度，它们和功率之间的关系是 $\varepsilon_c = \sqrt{2P_c/\hbar\omega_c}$，$\varepsilon_p = \sqrt{2P_p/\hbar\omega_p}$。$\phi_p$ 是探测场和控制场之间的相位差，并且我们假定机械驱动场的频率等于探测场与控制场之间的频率失谐量。忽略量子噪声和热噪声项，根据海森伯运动方程可以得到腔模和力学模式算符的时间演化方程：

$$\frac{d\hat{a}}{dt} = -\left[\kappa/2 + i\left(\Delta_0 + \sum_{k=1}^{N}G_k\hat{x}_k\right)\right]\hat{a} + \sqrt{\eta_c\kappa/2}(\varepsilon_c + \varepsilon_p e^{-i\Omega t - i\phi_p})$$

(4.1.3)

$$\frac{d\hat{x}_k}{dt} = \frac{\hat{p}_k}{m_k}$$

(4.1.4)

$$\frac{d\hat{p}_k}{dt} = -m_k\omega_k^2\hat{x}_k - \hbar G_k\hat{a}^+\hat{a} + 2\varepsilon_m\cos(\Omega t + \phi_m) - \gamma_k\hat{p}_k \quad (4.1.5)$$

其中腔的衰减率 κ 和机械振子的衰减率 γ_k 被唯象地引入。令方程(4.1.3)～方程(4.1.5)中时间求导项为零，可以得到腔场和机械振子位移的稳态解，满足以下的代数方程：

$$\bar{a} = \frac{\sqrt{\eta_c\kappa/2}}{\kappa/2 + i\bar{\Delta}}\varepsilon_c$$

(4.1.6)

$$m_k\omega_k^2\bar{x}_k + \hbar G_k|\bar{a}|^2 = 0$$

(4.1.7)

其中，$\bar{\Delta} = \Delta_0 + \sum_{k=1}^{N}G_k\bar{x}_k$ 是包括辐射压效应后的有效腔失谐量。

因为探测场比控制场弱很多，我们可以将每一个海森伯算符写成其稳态平均值和小的涨落的和的形式，即 $\hat{a}(t) = \bar{a} + \delta\hat{a}(t)$，$\hat{x}_k(t) = \bar{x}_k + \delta\hat{x}_k(t)$。接下来，只保留 $\delta\hat{a}$、$\delta\hat{a}^+$、$\delta\hat{x}_k$ 等小量的一阶项，可以得到以下线性化的海森伯-朗之万方程：

$$\frac{\mathrm{d}}{\mathrm{d}t}\delta\hat{a}(t) = -(\kappa/2 + \mathrm{i}\bar{\Delta})\delta\hat{a}(t) - \mathrm{i}\sum_{k=1}^{N} G_k \bar{a} \delta\hat{x}_k(t) + \sqrt{\eta_c\kappa/2}\,\varepsilon_p \mathrm{e}^{-\mathrm{i}(\Omega t + \phi_p)}$$

(4.1.8)

$$\frac{\mathrm{d}^2}{\mathrm{d}t^2}\delta\hat{x}_k(t) + \gamma_k \frac{\mathrm{d}}{\mathrm{d}t}\delta\hat{x}_k(t) + \omega_k^2 \delta\hat{x}_k(t)$$

$$= -\frac{\hbar G_k}{m_k}\bar{a}[\delta\hat{a}(t) + \delta\hat{a}^+(t)] + \frac{2\varepsilon_m}{m_k}\cos(\Omega t + \phi_m) \quad (4.1.9)$$

其中被忽略掉的 $\delta\hat{x}_k \delta\hat{a}$ 和 $\delta\hat{a}^+ \delta\hat{a}$ 等非线性项可以引起二阶边带[8,27]。需要指出的是,系统只有当线性化的朗之万方程(4.1.8)和方程(4.1.9)对应的本征值的实部全部为负值时才是稳定的。稳定性条件可以根据 Routh-Hurwitz 标准求出[43],但是具体的形式太复杂,这里就不给出具体表达式,我们将数值上验证本书中选取的参数满足稳定性条件。因为驱动场是经典的相干场,我们将把所有的算符用其期待值表示,即 $\langle\delta\hat{a}(t)\rangle = \delta a(t)$,$\langle\delta\hat{x}_k(t)\rangle = \delta x_k(t)$。为了求解方程(4.1.8)和方程(4.1.9),引入代换 $\delta a(t) = A^- \mathrm{e}^{-\mathrm{i}\Omega t} + A^+ \mathrm{e}^{\mathrm{i}\Omega t}$,$\delta x_k(t) = X_k \mathrm{e}^{-\mathrm{i}\Omega t} + X_k^* \mathrm{e}^{\mathrm{i}\Omega t}$。本书讨论的是可分辨边带区域($\kappa \ll \omega_k$)并且腔场被驱动至蓝边带($\bar{\Delta} = -\omega_m$),其中 ω_m 是 N 个机械振子的平均频率。在这种情况下,较低的边带 A^+ 远离共振,因此可以被忽略。将上述代换代入到方程(4.1.8)和方程(4.1.9),可以得到

$$A^- = \frac{\sqrt{\eta_c\kappa/2}\,\varepsilon_p \mathrm{e}^{-\mathrm{i}\phi_p}}{\kappa/2 - \mathrm{i}(\Omega + \omega_m) + \sum_{k}^{N} \hbar G_k^2 |\bar{a}|^2 \chi(m_k,\omega_k)} +$$

$$\frac{\sum_{k}^{N} \hbar G_k \bar{a} \chi(m_k,\omega_k) \varepsilon_m \mathrm{e}^{-\mathrm{i}\phi_m}}{\kappa/2 - \mathrm{i}(\Omega + \omega_m) + \sum_{k}^{N} \hbar G_k^2 |\bar{a}|^2 \chi(m_k,\omega_k)} \quad (4.1.10)$$

式中,

$$\chi(m_k,\omega_k) = \frac{1}{2m_k\omega_k[-\gamma_k/2 + \mathrm{i}(\Omega + \omega_k)]} \quad (4.1.11)$$

方程(4.1.10)右侧的第一项是探测场和控制场的贡献,可以产生通常的光力诱导吸收和参量放大,第二项是由机械驱动引起的声子-光子序参量过程。

光力腔的输出场可以根据输入-输出关系得到

$$\begin{aligned}a_{\text{out}}(t) &= a_{\text{in}}(t) - \sqrt{\eta_c \kappa/2}\, a(t) \\&= (\varepsilon_c - \sqrt{\eta_c \kappa/2}\,\bar{a})\mathrm{e}^{-\mathrm{i}\omega_c t} + (\varepsilon_p \mathrm{e}^{-\mathrm{i}\phi_p} - \sqrt{\eta_c \kappa/2}\, A^-)\mathrm{e}^{-\mathrm{i}\omega_p t} - \\&\quad \sqrt{\eta_c \kappa/2}\, A^+ \mathrm{e}^{-\mathrm{i}(2\omega_c - \omega_p)t}\end{aligned} \quad (4.1.12)$$

定义探测场透射谱为 $t = (\varepsilon_p \mathrm{e}^{-\mathrm{i}\phi_p} - \sqrt{\eta_c \kappa/2}\, A^-)/(\varepsilon_p \mathrm{e}^{-\mathrm{i}\omega_p t})$,可以得到

$$t = t_1 + t_2 \quad (4.1.13)$$

式中,

$$t_1 = 1 - \frac{\eta_c \kappa/2}{\kappa/2 - \mathrm{i}(\Omega + \omega_m) + \sum_k^N \hbar G_k^2 |\bar{a}|^2 \chi(m_k, \omega_k)} \quad (4.1.14)$$

$$t_2 = \frac{\sqrt{\eta_c \kappa/2} \sum_k^N G_k \bar{a} \chi(m_k, \omega_k) \varepsilon_m/\varepsilon_p \mathrm{e}^{-\mathrm{i}\phi}}{\kappa/2 - \mathrm{i}(\Omega + \omega_m) + \sum_k^N \hbar G_k^2 |\bar{a}|^2 \chi(m_k, \omega_k)} \quad (4.1.15)$$

其中,$\phi = \phi_m - \phi_p$ 是两种驱动场之间的相位差。t_1 是没有机械驱动时探测场透射谱的表达式,已经在之前包含单个机械振子的光力系统中讨论过[12]。t_2 表示的是由机械驱动场引起的对探测场透射谱的修正项。t_1 和 t_2 之间的干涉决定探测场透射谱,其中控制场功率 P_c、相位差 ϕ 以及机械驱动场的强度 ε_m 起着重要的作用。

4.1.3 结果和讨论

本节我们将数值上研究 N 个机械振子耦合于一个共同的机械振子探测场的透射谱。为了简单起见,首先讨论两个频率稍有不同的机械振子耦合于一个腔场构成的光力学系统的情况。根据最近的实验选取以下参数[30]:$\omega_0 = 2\pi \times 6.98\text{ GHz}, \omega_1 = 2\pi \times 32.1\text{ MHz}, \omega_2 = 2\pi \times 32.5\text{ MHz}, \kappa = 2\pi \times 6.2\text{ MHz}, m_1 = 557\text{ fg}, m_2 = 534\text{ fg}, \gamma_1 = \gamma_2 = \gamma_m = 2\pi \times 930\text{ Hz}, G_1 = 2\pi \times 1.8\text{ MHz/nm}, G_2 = 2\pi \times 2\text{ MHz/nm}$。这里我们只讨论腔场被驱动至其蓝边带的情况,即 $\bar{\Delta} = -\omega_m$。

在没有机械驱动的情况下,图 4.1.2 所示是 $P_c = 2\text{ nW}$ 时探测场透射谱 $|t|^2$ 随 $(\Omega + \omega_1)/(2\pi)$ 变化的情况。可以看到图的中央有两个非常窄的吸收

凹陷,插图清楚地显示这两个凹陷分别位于 $(\Omega+\omega_1)/(2\pi)=0$ 和 $(\Omega+\omega_1)/(2\pi)=-0.4\,\mathrm{MHz}$。根据所选的机械振子的共振频率,可以发现探测场透射谱中两个吸收凹陷分别位于 $\Omega=-\omega_1$ 和 $\Omega=-\omega_2$,被称为双光力诱导吸收。与其他光力系统中的单光力诱导吸收类似,这种现象可由振动频率为探测场和控制场之间拍频的辐射压力解释。当拍频与机械振子的共振频率接近时,即 $\Omega=-\omega_k(k=1,2)$,机械振子开始相干振荡,从而产生强控制场的斯托克斯和反斯托克斯散射。当腔场被驱动至其蓝边带 $(\omega_0-\omega_c\approx-\omega_m)$ 且系统工作在可分辨边带区域 $(\omega_1>\kappa,\omega_2>\kappa)$ 时,频率为 $\omega_c+\omega_k$ 的反斯托克斯场因为与腔场非共振而被强烈抑制,腔内只有频率为 $\omega_c-\omega_k$ 的斯托克斯场。此外,斯托克斯场与简并的探测场之间的相长干涉增强了腔内的探测场。腔对探测场光子的增强吸收表现为探测场透射谱中 $\Omega=-\omega_1$ 和 $\Omega=-\omega_2$ 两处的透射率变得更小。值得指出的是,当两个机械振子的共振频率相同并且没有直接的相互作用时,探测场透射谱中只有一个吸收凹陷。

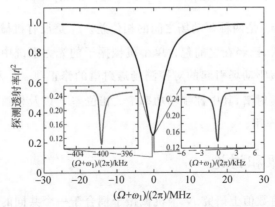

图 4.1.2　没有机械驱动且腔场被驱动至蓝边带时探测场透射谱 $|t|^2$ 随探测失谐量 $(\Omega+\omega_1)/(2\pi)$ 变化的曲线。插图表明在 $\Omega=-\omega_1$ 和 $\Omega=-\omega_2$ 处有两个尖锐的吸收凹陷

(请扫Ⅱ页二维码看彩图)

接下来,我们将主要研究机械驱动对探测场透射谱的影响。简单起见,我们假设每个机械振子分别受到强度为 ε_m、频率为 Ω、相位为 ϕ_m 的相干驱动场驱动。图 4.1.2 表明探测场透射谱中有两个吸收凹陷。不失一般性,我们只需要研究其中一个吸收凹陷,另外一个应该类似。需要指出的是,两个机械振子之间的频率差应该大于机械振子的衰减率 γ_m,这样透射谱中的两个吸

收凹陷才不会重叠。图 4.1.3 所示是 $P_c = 2\text{ nW}$ 时 $\Omega = -\omega_1$ 处探测场的峰值透射率(即 $|t_p|^2 = |t|^2_{\Omega=-\omega_1}$)随外界驱动力的相位差 ϕ 和强度 ε_m 变化的情况。可以看出,相位差 ϕ 对峰值透射率 $|t_p|^2$ 的影响随着驱动强度的增加变得更加明显。对于某一给定的驱动强度 ε_m,$\Omega = -\omega_1$ 处的 $|t_p|^2$ 在 $\phi = \pi/2$ 时有最大值而在 $\phi = 3\pi/2$ 时有最小值,可以通过干涉过程进行解释。此外,相位差处于一定范围时,当驱动强度 ε_m 大于某一临界值时探测场峰值透射率可大于 1。因此,通过控制机械驱动力的相位和强度可使该系统实现从双光力诱导吸收到参量放大的转变。

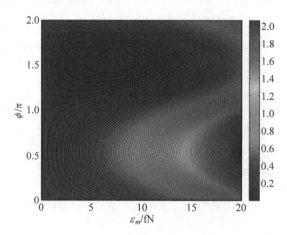

图 4.1.3 $\Omega = -\omega_1$ 处探测场的峰值透射率 $|t_p|^2$ 随相位差 ϕ 和机械强度 ε_m 的变化图

(请扫 II 页二维码看彩图)

为了更清楚地看到机械驱动的影响,图 4.1.4 中给出了不同机械驱动下探测场透射谱随探测场失谐量 $(\Omega + \omega_1)/(2\pi)$ 变化的情况。如图 4.1.4(a)所示,没有机械驱动时($\varepsilon_m = 0$),在 $\Omega = -\omega_1$ 处透射率 $|t|^2$ 有极小值。当机械驱动力的大小为 $\varepsilon_m = 12\text{ fN}$,相位差 $\phi = \pi/2$ 时探测场透射被大大增强,而相位差 $\phi = 3\pi/2$ 时探测场透射率反而进一步减小。因此,相位依赖效应非常明显。图 4.1.4(b)~(d)所示分别是 $\phi = 0$、$\phi = \pi/2$、$\phi = 3\pi/2$ 时不同机械驱动强度下探测场透射谱随 $(\Omega + \omega_1)/(2\pi)$ 变化的情况。当 $\phi = 0$ 和 $\phi = \pi/2$ 时,$\Omega = -\omega_1$ 处的探测场峰值透射率随着机械驱动强度的增强而单调增加。但

是,当 $\phi=3\pi/2$ 时,机械驱动强度的增加并不一定会引起峰值透射率的增大。当驱动强度比较小时,如 $\varepsilon_m \leqslant 7$ fN,峰值透射率随着 ε_m 的增加而减小,但是当 ε_m 进一步增大到 20 fN 时,峰值透射率又进一步增强。这种相位依赖的效应使得包含机械驱动的光力系统中的探测场透射谱更具有可调性。

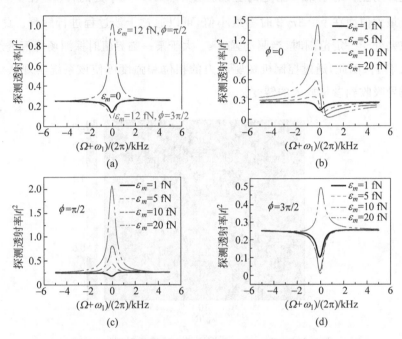

图 4.1.4 不同机械驱动强度 ε_m 和相位差 ϕ 下探测场透射谱随探测失谐量 $(\Omega+\omega_1)/(2\pi)$ 变化的曲线

(请扫 II 页二维码看彩图)

为了能够解释上述相位依赖的效应,我们在图 4.1.5 中给出了在 $\Omega=-\omega_1$ 处 $|t_p|^2$、$|t_{1p}|^2$ 和 $|t_{2p}|^2$ 随机械驱动力 ε_m 变化的情况。随着 ε_m 的增大,$|t_{1p}|^2$ 保持不变但 $|t_{2p}|^2$ 单调增加。当 $\phi=\pi/2$ 时,t_{1p} 和 t_{2p} 之间产生相长干涉。如果它们之间的大小相等,干涉效应最明显,导致在 $\varepsilon_m=7$ fN 附近 $|t_p|^2 \approx 4|t_{1p}|^2 = 4|t_{2p}|^2$,如图 4.1.5(a)所示。当 $\phi=3\pi/2$ 时,t_{1p} 和 t_{2p} 之间产生相消干涉,从图 4.1.5(b)可以看出 $\varepsilon_m=7$ fN 时 $|t_p|^2 \approx 0.0023$,这时相消干涉最彻底,光子的能量被转移到了机械振子之中。此外,图 4.1.5(a)和(b)表明,t_{1p} 和 t_{2p} 之间的大小相差越多,干涉效应越弱。

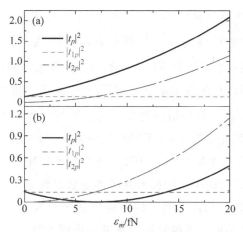

图 4.1.5　(a) $\phi=\pi/2$ 和 (b) $\phi=3\pi/2$ 时 $|t_p|^2$、$|t_{1p}|^2$、$|t_{2p}|^2$ 随机械驱动强度 ε_m 变化的曲线

(请扫Ⅱ页二维码看彩图)

在上述讨论中，控制场始终保持恒定，即 $\bar{\Delta}=-\omega_m$，$P_c=2$ nW。图 4.1.6 给出了 $\Omega=-\omega_1$ 处的峰值透射率 $|t_p|^2$ 在不同机械驱动下随控制场功率变化的情况。在这四种情况下，控制场功率的影响是类似的。首先，随着控制功率的增大峰值透射率从功率为零时的初始值逐渐降低到一个最小值，但是进一步增加功率会使 $|t_p|^2$ 又开始变大。当功率足够强时，$|t_p|^2$ 甚至可以大于 1，这意味着系统进入了序参量放大区域。当机械驱动的强度保

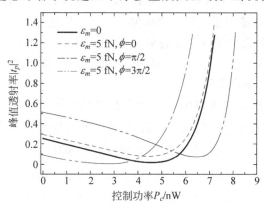

图 4.1.6　峰值透射谱 $\Omega=-\omega_1$ 处的峰值透射率 $|t_p|^2$ 在不同机械驱动下随控制场功率变化的情况

(请扫Ⅱ页二维码看彩图)

持一定时,峰值透射率随控制场功率变化时的相位依赖关系很明显。$\varepsilon_m=5$ fN,$\phi=0$ 时的曲线与没有机械驱动时的曲线类似,但是 $\phi=\pi/2$ 和 $\phi=3\pi/2$ 两条曲线有着明显的不同。我们对这些现象的物理解释如下:辐射压和机械驱动都可以导致机械振子的相干振动,从而产生两种不同的斯托克斯散射。相位差 ϕ 在这两种斯托克斯场与探测场之间的干涉中起着非常重要的作用。此外,斯托克斯光子(下转换的控制光子)的多少取决于控制场的功率。如果控制场的功率 P_c 非常低,下转换的控制光子数低于入射的探测光子数。因此,干涉效应引起的探测光子的受激吸收是不完全的。随着控制功率的增加,吸收凹陷加深。当下转换的控制光子数等于探测光子数时,探测场透射率出现最小值。当控制场功率进一步增加时,下转换的控制光子数大于探测光子数,导致探测场透射率又开始变大。当控制场功率大于某一临界值时,大小取决于系统参数以及机械驱动的相位和强度,探测场透射率可以大于 $1^{[12]}$。需要注意的是,因为始终保持 $\varepsilon_p=\varepsilon_c/1000$,随着控制场功率的增加探测场功率也随之增加,但是 ε_m 并未改变,因此方程(4.1.10)等号右边两项的贡献也随着控制场功率而改变,并且 $\phi=\pi/2$ 和 $\phi=3\pi/2$ 两条曲线之间有一个交点。

接下来,我们考虑由三个不同的机械振子耦合于一个共同的腔场形成的多模光力系统。基于已有的实验参数[30],我们假定第三个机械振子的参数为: $\omega_3=2\pi\times31.7$ MHz,$m_3=580$ fg,$G_3=2\pi\times1.6$ MHz/nm,$\gamma_3=2\pi\times930$ Hz。图 4.1.7 给出了机械驱动施加前后探测场透射率 $|t|^2$ 随探测失谐量 $(\Omega+\omega_1)/(2\pi)$ 变化的情况,在透射谱的下端有三个吸收凹陷,分别位于 $\Omega=-\omega_1$,$\Omega=-\omega_2$ 和 $\Omega=-\omega_3$。此外,机械驱动的影响与两个机械振子时类似,如图 4.1.7 的插图所示。

上述研究表明在包含两个或三个非简并机械振子的光力系统中可以出现双色或三色光力诱导透明现象。最后,我们考虑包含 N 个频率稍有不同的机械振子的光力系统,定性的分析可以发现在探测场透射谱中 $\Omega=-\omega_k(k=1,2,3,\cdots,N)$ 的位置可以最多出现 N 个吸收凹陷。物理原因也可以通过探测场和频率为 $\omega_c-\omega_k(k=1,2,3,\cdots,N)$ 的斯托克斯场之间的干涉效应进行解释。如果 $L(1<L<N)$ 个机械振子有共同的频率 $\omega(\omega_1=\omega_2=\cdots=\omega)$,并且剩下的 $N-L$ 的机械振子的频率不是 ω 且各不相同,则在探测场透射谱中有 $N-L+1$ 个吸收凹陷。但是,如果所有的机械振子都有相同的共振频率

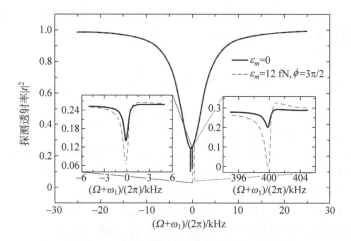

图 4.1.7 考虑机械驱动时探测场透射率 $|t|^2$ 随探测失谐量 $(\Omega+\omega_1)/(2\pi)$ 变化的图。插图表明在 $\Omega=-\omega_1$ 和 $\Omega=-\omega_3$ 处有两个尖锐的吸收凹陷。三个机械振子的参数为：$\omega_3=2\pi\times31.7\text{ MHz}, m_3=580\text{ fg}$, $G_3=2\pi\times1.6\text{ MHz/nm}, \gamma_3=2\pi\times930\text{ Hz}$

（请扫Ⅱ页二维码看彩图）

$\omega_k=\omega_m$，则只有一个吸收凹陷出现在 $\Omega=-\omega_m$ 处。实验上，可以通过微波领域的光力相互作用[30]或者将多个不同的原子系综受限于一个共同的光学腔[44]实现多个机械振子耦合于一个共同腔的结构。值得一提的是，多模光力系统中的集体效应最近受到越来越多的研究。比如，基普夫(Kipfh)和阿加瓦尔(Agarwal)等研究发现，通过调节泵浦场的频率失谐量可以使系统从超辐射区域转变成集体增益区域[45]。张(Zhang)等从实验上证实了机械振子序列耦合于一个共同的光学腔时可以实现同步振动[46]。

4.1.4 小结

综上所述，我们理论上研究了包含 N 个近简并的机械振子的光力系统在强的控制场和弱的机械驱动场共同作用下对弱的探测场的响应[47]。当腔场被驱动至其蓝边带，我们得到了探测场透射谱的解析表达式。我们发现该系统探测场透射谱中最多可出现 N 个窄的吸收凹陷，即多色光力诱导吸收现象，并且可以通过机械驱动场的强度和相位进行有效调制。控制场功率较低时，机械驱动的强度和相位可以用于实现系统中光力诱导吸收和参量放大之间的转变。此外,控制场功率较高时仍然可以出现弱的探测场的参量放大现象,其中相位依赖的效应非常明显。

参 考 文 献

[1] KIPPENBERG T J, VAHALA K J. Cavity optomechanics: back-action at the mesoscale[J]. Science, 2008, 321: 1172.
[2] MARQUARDT F, GIRVIN S M. Optomechanics[J]. Physics, 2009, 2: 40.
[3] ASPELMEYER M, KIPPENBERG T J, MARQUARDT F. Cavity optomechanics [J]. Rev. Mod. Phys. , 2014, 86: 1391.
[4] AGARWAL G S, HUANG S. Electromagnetically induced transparency in mechanical effects of light[J]. Phys. Rev. A, 2010, 81: 041803.
[5] WEIS S, RIVIÈRE R, DELÉGLISE S, et al. Optomechanically induced transparency [J]. Science, 2010, 330: 1520.
[6] SAFAVI-NAEINI A H, ALEGRE T P M, CHAN J, et al. Electromagnetically induced transparency and slow light with optomechanics[J]. Nature, 2011, 472: 69-73.
[7] TEUFEL J D, LI D, ALLMAN M S, et al. Circuit cavity electromechanics in the strong-coupling regime[J]. Nature, 2011, 471: 204.
[8] XIONG H, SI L G, ZHENG A S, et al. Higher-order sidebands in optomechanically induced transparency[J]. Phys. Rev. A, 2012, 86: 013815.
[9] KARUZA M, BIANCOFIORE C, BAWAJ M, et al. Optomechanically induced transparency in a membrane-in-the-middle setup at room temperature[J]. Phys. Rev. A, 2013, 88: 013804.
[10] KRONWALD A, MARQUARDT F. Optomechanically induced transparency in the nonlinear quantum regime[J]. Phys. Rev. Lett. , 2013, 111: 133601.
[11] JING H, ÖZDEMIR S K, GENG Z, et al. Optomechanically-induced transparency in parity-time-symmetric microresonators[J]. Sci. Rep. , 2015, 5: 9663.
[12] HOCKE F, ZHOU X, SCHLIESSER A, et al. Electromechanically induced absorption in a circuit nano-electromechanical system[J]. New J. Phys. , 2012, 14: 123037.
[13] QU K N, AGARWAL G S. Phonon-mediated electromagnetically induced absorption in hybrid opto-electromechanical systems[J]. Phys. Rev. A, 2013, 87: 031802.
[14] SINGH V, BOSMAN S J, SCHNEIDER B H, et al. Optomechanical coupling between a multilayer graphene mechanical resonator and a superconducting microwave cavity[J]. Nat. Nanotechnol. , 2014, 9: 820.

[15] MASSEL F, HEIKKILÄ T T, PIRKKALAINEN J M, et al. Microwave amplification with nanomechanical resonators[J]. Nature,2011,480: 351.

[16] FLEISCHHAUER M, IMAMOGLU A, MARANGOS J P. Electromagnetically induced transparency: optics in coherent media[J]. Rev. Mod. Phys. , 2005, 77: 633.

[17] WU Y, YANG X X. Electromagnetically induced transparency in V-, Λ-, and cascade-type schemes beyond steady-state analysis[J]. Phys. Rev. A, 2005, 71: 053806.

[18] AKULSHIN A M, BARREIRO S, LEZAMA A. Electromagnetically induced absorption and transparency due to resonant two-field excitation of quasidegenerate levels in Rb vapor[J]. Phys. Rev. A,1998,57: 2996.

[19] JIANG C, LIU H X, CUI Y S, et al. Electromagnetically induced transparency and slow light in two-mode optomechanics[J]. Opt. Express,2013,21: 12165.

[20] ZHOU X, HOCKE F, SCHLIESSER A, et al. Slowing, advancing and switching of microwave signals using circuit nanoelectromechanics[J]. Nat. Phys. , 2013, 9: 179.

[21] CHANG D E, SAFAVI-NAEINI A H, HAFEZI M, et al. Slowing and stopping light using an optomechanical crystal array[J]. New J. Phys. ,2011,13: 023003.

[22] FIORE V, YANG Y, KUZYK M C, et al. Storing optical information as a mechanical excitation in a silica optomechanical resonator[J]. Phys. Rev. Lett. , 2011,107: 133601.

[23] SCULLY M O, ZHU S Y, GAVRIELIDES A. Degenerate quantum-beat laser: lasing without inversion and inversion without lasing[J]. Phys. Rev. Lett. ,1989, 62: 2813.

[24] JIA W Z, WEI L F, LI Y, et al. Phase-dependent optical response properties in an optomechanical system by coherently driving the mechanical resonator[J]. Phys. Rev. A,2015,91: 043843.

[25] MA J Y, YOU C, SI L G, et al. Optomechanically induced transparency in the presence of an external time-harmonic-driving force[J]. Sci. Rep. ,2015,5: 11278.

[26] XU X W, LI Y. Controllable optical output fields from an optomechanical system with mechanical driving[J]. Phys. Rev. A,2015,92: 023855.

[27] SUZUKI H, BROWN E, STERLING R. Nonlinear dynamics of an optomechanical system with a coherent mechanical pump: second-order sideband generationendnote [J]. Phys. Rev. A,2015,92: 033823.

[28] FAN L, FONG K Y, POOT M, et al. Cascaded optical transparency in multimode-cavity optomechanical systems[J]. Nat. Commun. ,2015,6: 5850.

[29] SEOK H, BUCHMANN L F, WRIGHT E M, et al. Multimode strong-coupling quantum optomechanics[J]. Phys. Rev. A,2013,88: 063850.

[30] MASSEL F, CHO S U, PIRKKALEINEN J M, et al. Multimode circuit

optomechanics near the quantum limit[J]. Nat. Commun. ,2012,3: 987.

[31] XU X W,LI Y,CHEN A X,et al. Nonreciprocal conversion between microwave and optical photons in electro-optomechanical systems [J]. Phys. Rev. A, 2016, 93: 023827.

[32] HILL J T, SAFAVI-NAEINI A H, CHAN J, et al. Coherent optical wavelength conversion via cavity optomechanics[J]. Nat. Commun. ,2012,3: 1196.

[33] LIU Y, DAVANCO M, AKSYUK V, et al. Electromagnetically induced transparency and wideband wavelength conversion in silicon nitride microdisk optomechanical resonators[J]. Phys. Rev. Lett. ,2013,110: 223603.

[34] ANDREWS R W,PETERSON R W,PURDY T P,et al. Bidirectional and efficient conversion between microwave and optical light[J]. Nat. Phys. ,2014,10: 321.

[35] DONG C, FIORE V, KUZYK M C, et al. Optical wavelength conversion via optomechanical coupling in a silica resonator[J]. Ann. Phys. ,2015,527: 100.

[36] LECOCQ F, CLARK J B, SIMMONDS R W, et al. Mechanically mediated microwave frequency conversion in the quantum regime[J]. Phys. Rev. Lett. , 2016,116: 043601.

[37] SHKARIN A B,FLOWERS-JACOBS N E,HOCH S W,et al. Optically mediated hybridization between two mechanical modes [J]. Phys. Rev. Lett. , 2014, 112: 013602.

[38] WOOLLEY M J,CLERK A A. Two-mode squeezed states in cavity optomechanics via engineering of a single reservoir[J]. Phys. Rev. A,2014,89: 063805.

[39] HUANG S M. Double electromagnetically induced transparency and narrowing of probe absorption in a ring cavity with nanomechanical mirrors[J]. J. Phys. B: At. Mol. Opt. Phys. ,2014,47: 055504.

[40] MA P C,ZHANG J Q,XIAO Y,et al. Tunable double optomechanically induced transparency in an optomechanical system[J]. Phys. Rev. A,2014,90: 043825.

[41] WANG H, GU X, LIU Y X, et al. Optomechanical analog of two-color electromagnetically induced transparency: photon transmission through an optomechanical device with a two-level system [J]. Phys. Rev. A, 2014, 90: 023817.

[42] BUCHMANN L F, STAMPER-KURN D M. Nondegenerate multimode optomechanics[J]. Phys. Rev. A,2015,92: 013851.

[43] DEJESUS E X, KAUFMAN C. Routh-Hurwitz criterion in the examination of eigenvalues of a system of nonlinear ordinary differential equations[J]. Phys. Rev. A,1987,35: 5288.

[44] BOTTER T, BROOKS D W C, SCHREPPLER S, et al. Optical readout of the quantum collective motion of an array of atomic ensembles[J]. Phys. Rev. Lett. , 2013,110: 153001.

[45] KIPF T, AGARWAL G S. Superradiance and collective gain in multimode

optomechanics[J]. Phys. Rev. A,2014,90: 053808.
[46] ZHANG M, SHAH S, CARDENAS J, et al. Synchronization and phase noise reduction in micromechanical oscillator arrays coupled through light[J]. Phys. Rev. Lett. ,2015,115: 163902.
[47] JIANG C,CUI Y S,BIAN X T,et al. Phase-dependent multiple optomechanically induced absorption in multimode optomechanical systems with mechanical driving [J]. Phys. Rev. A,2016,94: 023837.

4.2 混杂光力系统中可控的光学响应

4.2.1 引言

迅速发展的腔光力学领域研究电磁和机械系统通过辐射压形成的非线性相互作用,不但可以利用光实现对机械运动的量子调控,还可以利用光力相互作用控制光力系统的光学响应[1-4]。根据电磁场的频率,光学和微波频率的光力系统取得了十分重要的进展,包括纳米机械振子的量子基态冷却[5,6]、光和力学自由度之间的量子相干耦合[7,8]、正则模式分裂[9,10]、光力诱导透明[11-16]等。光力诱导透明类似于电磁诱导透明[17,18],而电磁诱导透明首先在原子气中被观察到[19],后来逐渐在各种固态系统中被观察到,如量子阱[20]、氮空位色心[21]。同样,光力诱导透明可以用来实现延迟或者超前电磁信号[13,22]以及将信息存储在寿命较长的机械运动中[23]。此外,蓝边带泵浦场驱动下实现的光力诱导吸收[24]和放大[25]现象在由超导微波腔和纳米机械振子构成的微波光力系统中被观察到。辛格等在由石墨烯和超导微波腔构成的光力系统中观察到了类似的现象[26]。

最近,实验上可以成功地将一个机械振子耦合于不同频率的电磁腔,这引起了对双腔光力系统的广泛研究。这样的系统可以用来实现两个不同光学波长[27-29]或者光学和微波频率[30,31]之间的相干转换。博赫曼(Bochmann)等利用压电光子晶体实现了一种光学光子和微波电信号之间的纳米机械界面,并且观察到了电力诱导的光学透明现象[32]。此外,通过将电力学和光力学结合起来,安德鲁斯(Andrews)等成功地将一个纳米机械振子同时耦合于一个超导微波腔和一个光学腔[33],从而形成了一种混杂光-电力系统[34,35]。他们实验上观察到了微波和光学光子之间可逆的相干转化,在现代通信网络中有着重要的应用。李(Li)等利用大失谐的驱动激光研究了该混杂光力系统中分束器和双模压缩光子-光子相互作用[36]。本节我们研究该混杂系统在两束强的泵浦场和一束弱的探测场同时作用下如何实现光力诱导吸收和放大之间的转换。值得指出的是,曲(Qu)和阿加瓦尔(Agarwal)等发现当该双腔光力系统的两个腔都驱动在各自的红边带时,透明窗口中会出现一个吸收峰[37]。但是观察到该现象的一个条件是光学腔的衰减率 κ_o 大于机械振子的衰减率 γ_m 又大于微波腔的衰减率 κ_e(即 $\kappa_o > \gamma_m > \kappa_e$),而这一条件目前在实验中还很难实现。最近,努尔内坎普(Nunnenkamp)等首次研究了机械振子衰减率大于电磁腔衰减率的情况,并发现可以实现力学诱导的电磁模式的放大[38]。本节我们根据实验上可行的参数研究了混杂光电系统中的光力诱导吸收[39]和放大现象,其中一个腔受到一束蓝失谐的泵浦场驱动,而另外一个腔受到一束红失谐的泵浦场驱动。因此,我们可以通过调节泵浦场的功率实现光力诱导吸收和放大之间的转换。该混杂系统可被用于实现量子极限的放大[40]。

4.2.2 模型和理论

混杂光力系统的示意图如图 4.2.1 所示,其中一个微波腔和一个光学腔同时耦合于一个共同的纳米机械振子。共振频率为 ω_o 的光学腔受到一束振幅为 E_o、频率为 Ω_o 的强泵浦场以及一束振幅为 E_p、频率为 Ω_p 的弱探测场同时驱动。共振频率为 ω_e 的微波腔可由等效电感 L 和等效电容 C 表示,且只受到振幅为 E_e、频率为 Ω_e 的强泵浦场驱动。在泵浦频率 Ω_o 和 Ω_e 的旋转框架下,混杂系统的哈密顿量表示为[12,30,31]

$$H = H_0 + H_{int} + H_{drive}$$

式中,

$$H_0 = \hbar \Delta_o a^\dagger a + \hbar \Delta_e b^\dagger b + \hbar \omega_m c^\dagger c$$
$$H_{int} = -\hbar g_0 a^+ a (c^+ + c) - \hbar g_e b^+ b (c^+ + c)$$
$$H_{drive} = i\hbar \sqrt{\kappa_{o,ext}/2} E_o (a^+ - a) + i\hbar \sqrt{\kappa_{e,ext}/2} E_e (b^+ - b) +$$
$$i\hbar \sqrt{\kappa_{o,ext}/2} E_p (a^+ e^{-i\delta t} - a e^{i\delta t})$$

(4.2.1)

式中,算符 a、b 和 c 分别是光学腔、微波腔和机械振子的湮灭算符。$\Delta_o =$

$\omega_o - \Omega_o$ 和 $\Delta_e = \omega_e - \Omega_e$ 分别是腔泵浦场的失谐量。ω_m 是衰减率为 γ_m 的机械振子的共振频率。$g_o(g_e)$ 是机械振子和光学腔(微波腔)之间的单光子耦合强度。H_{drive} 表示输入场和腔场之间的相互作用,其中 E_o、E_e、E_p 和各自的功率之间的关系式为 $|E_o| = \sqrt{2P_o/\hbar\Omega_o}$,$|E_e| = \sqrt{2P_e/\hbar\Omega_e}$,$|E_p| = \sqrt{2P_p/\hbar\Omega_p}$[41]。$\kappa_o(\kappa_e)$ 是光学腔(微波腔)总的衰减率,$\kappa_{o,\text{ext}}(\kappa_{e,\text{ext}})$ 是由外部耦合引起的衰减率[33]。$\delta = \Omega_p - \Omega_o$ 是探测场与泵浦场之间的失谐量。

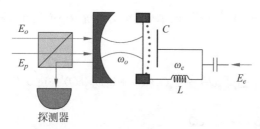

图 4.2.1 混杂光力系统示意图。机械振子同时耦合于共振频率为 ω_o 的光学腔和共振频率为 ω_e 的微波腔,微波腔由等效电感 L 和等效电容 C 表示。光学腔同时受到振幅为 E_o 的强泵浦场和振幅为 E_p 的弱探测场驱动,而微波腔只受到振幅为 E_e 的强泵浦场驱动

(请扫Ⅱ页二维码看彩图)

根据海森伯运动方程及对易关系 $[a,a^+]=1$,$[b,b^+]=1$,$[c,c^+]=1$,可以得到算符 a、b、$Q(Q=c^++c)$ 的时间演化方程。此外,通过引入腔模和机械振子相应的噪声和衰减项,可以得到以下的量子朗之万方程:

$$\dot{a} = -\mathrm{i}(\Delta_o - g_o Q)a - \frac{\kappa_o}{2}a + \sqrt{\kappa_{o,\text{ext}}/2}(E_o + E_p \mathrm{e}^{-\mathrm{i}\delta t}) + \sqrt{\kappa_o}\, a_{\text{in}} \tag{4.2.2}$$

$$\dot{b} = -\mathrm{i}(\Delta_e - g_e Q)b - \frac{\kappa_e}{2}b + \sqrt{\kappa_{e,\text{ext}}/2}\, E_e + \sqrt{\kappa_e}\, b_{\text{in}} \tag{4.2.3}$$

$$\ddot{Q} + \gamma_m \dot{Q} + \omega_m^2 Q = 2g_o \omega_m a^+ a + 2g_e \omega_m b^+ b + \xi \tag{4.2.4}$$

式中,a_{in} 和 b_{in} 是平均值为零的输入真空噪声,ξ 是平均值为零的布朗随机力[42]。通过令方程(4.2.2)~方程(4.2.4)的时间求导项为零,可以得到如下的稳态解:

$$a_s = \frac{\sqrt{\kappa_{o,\text{ext}}/2} E_o}{\kappa_o/2 + i\Delta_o'}, \quad b_s = \frac{\sqrt{\kappa_{e,\text{ext}}/2} E_e}{\kappa_e/2 + i\Delta_e'}, \quad Q_s = \frac{2}{\omega_m}(g_o |a_s|^2 + g_e |b_s|^2)$$
(4.2.5)

式中,$\Delta_o' = \Delta_o - g_o Q_s$,$\Delta_e' = \Delta_e - g_e Q_s$ 是包含辐射压效应后的有效腔失谐量。接下来,通过将每个海森伯算符写成稳态值加小的涨落的形式(即 $a = a_s + \delta a$,$b = b_s + \delta b$,$Q = Q_s + \delta Q$)可以得到以下线性化的海森伯-朗之万方程:

$$\langle \delta \dot{a} \rangle = -\left(\frac{\kappa_o}{2} + i\Delta_o\right)\langle \delta a \rangle + ig_o Q_s \langle \delta a \rangle + ig_o a_s \langle \delta Q \rangle + \sqrt{\kappa_{o,\text{ext}}/2} E_p e^{-i\delta t}$$
(4.2.6)

$$\langle \delta \dot{b} \rangle = -\left(\frac{\kappa_e}{2} + i\Delta_e\right)\langle \delta b \rangle + ig_e Q_s \langle \delta b \rangle + ig_e b_s \langle \delta Q \rangle + ig_e b_s \langle \delta Q \rangle \quad (4.2.7)$$

$$\langle \delta \ddot{Q} \rangle + \gamma_m \langle \delta \dot{Q} \rangle + \omega_m^2 \langle \delta Q \rangle$$
$$= 2\omega_m g_o a_s (\langle \delta a \rangle + \langle \delta a^+ \rangle) + 2\omega_m g_e b_s (\langle \delta b \rangle + \langle \delta b^+ \rangle) \quad (4.2.8)$$

其中我们将所有的算符用各自的期待值代替并忽略了量子和热噪声项[12]。$\delta a^+ \delta a$、$\delta b^+ \delta b$、$\delta a \delta Q$、$\delta b \delta Q$ 等非线性项可以引起光力系统中一些有趣的现象,比如二阶和高阶边带等[43]。当泵浦场功率足够大时,$|a_s| \gg 1$,$|b_s| \gg 1$。因此,线性化方程(4.2.6)~方程(4.2.8)中的非线性项可以忽略。为了求解方程(4.2.6)~方程(4.2.8),作代换[44] $\langle \delta a \rangle = a_+ e^{-i\delta t} + a_- e^{i\delta t}$,$\langle \delta b \rangle = b_+ e^{-i\delta t} + b_- e^{i\delta t}$,$\langle \delta Q \rangle = Q_+ e^{-i\delta t} + Q_- e^{i\delta t}$,然后代入方程组可以得到以下解:

$$a_+ = \frac{\sqrt{\kappa_{o,\text{ext}}/2} E_p}{\kappa_o/2 + i\Delta_o' - i\delta} - \frac{ig_o^2 n_o}{f(\delta)} \frac{\sqrt{\kappa_{o,\text{ext}}/2} E_p}{(\kappa_o/2 + i\Delta_o' - i\delta)^2} \quad (4.2.9)$$

式中,

$$f(\delta) = \sum_{k=o,e} \frac{2\Delta_k' g_k^2 n_k}{(\kappa_k/2 - i\delta)^2 + \Delta_k'^2} - \frac{\omega_m^2 - \delta^2 - i\delta\gamma_m}{\omega_m} \quad (4.2.10)$$

腔内光子数 $n_o = |a_s|^2$,$n_e = |b_s|^2$ 由以下耦合方程决定:

$$n_o = \frac{\kappa_{o,\text{ext}}/2 E_o^2}{\kappa_o^2/2 + [\Delta_o - 2g_o/\omega_m(g_o n_o + g_e n_e)]^2} \quad (4.2.11)$$

$$n_e = \frac{\kappa_{e,\text{ext}}/2 E_e^2}{\kappa_e^2/2 + [\Delta_e - 2g_e/\omega_m(g_o n_o + g_e n_e)]^2} \quad (4.2.12)$$

腔的输出场可以通过标准的输入-输出理论得出[12,45]:$a_{\text{out}}(t) = a_{\text{in}}(t) - $

$\sqrt{\kappa_{o,\text{ex}}/2}\,a(t)$,其中 $a_{\text{out}}(t)$ 是输出场算符。考虑光学腔的输出场时,可以得到

$$\begin{aligned}a_{\text{out}}(t) &= (E_o - \sqrt{\kappa_{o,\text{ext}}/2}\,a_s)e^{-i\Omega_o t} + (E_p - \sqrt{\kappa_{o,\text{ext}}/2}\,a_+)e^{-i(\delta+\Omega_o)t} - \\ &\quad \sqrt{\kappa_{o,\text{ext}}/2}\,a_- e^{-i(\delta-\Omega_o)t} \\ &= (E_o - \sqrt{\kappa_{o,\text{ext}}/2}\,a_s)e^{-i\Omega_o t} + (E_p - \sqrt{\kappa_{o,\text{ext}}/2}\,a_+)e^{-i\Omega_p t} - \\ &\quad \sqrt{\kappa_{o,\text{ext}}/2}\,a_- e^{-i(2\Omega_o-\Omega_p)t}\end{aligned} \quad (4.2.13)$$

从方程(4.2.13)可以看出,输出场中包含两个输入分量(Ω_o 和 Ω_p)以及一个新产生的频率为 $2\Omega_o - \Omega_p$ 的四波混频分量。探测场的透射系数定义为输出场和输入场中探测场频率分量的振幅之比[12],具体形式为

$$t(\Omega_p) = \frac{E_p - \sqrt{\kappa_{o,\text{ext}}/2}\,a_+}{E_p} = 1 - \left[\frac{\kappa_{o,\text{ext}}/2}{\kappa_o/2 + i\Delta'_o - i\delta} - \frac{1}{f(\delta)}\frac{ig_o^2 n_o \kappa_{o,\text{ext}}/2}{(\kappa_o/2 + i\Delta'_o - i\delta)^2}\right] \quad (4.2.14)$$

4.2.3 数值结果和讨论

为了展示数值结果,我们使用实验上实现的混杂光力系统的参数[5,14,33]: $\omega_o = 2\pi \times 282$ THz, $\omega_e = 2\pi \times 7.1$ GHz, $\kappa_o = 2\pi \times 1.65$ MHz, $\kappa_e = 2\pi \times 1.6$ MHz, $\kappa_{o,\text{ext}} = 0.76\kappa_o$, $\kappa_{e,\text{ext}} = 0.11\kappa_e$, $g_o = 2\pi \times 27$ Hz, $g_e = 2\pi \times 2.7$ Hz, $\omega_m = 2\pi \times 5.6$ MHz, $\gamma_m = 2\pi \times 4$ Hz。

该系统的光学响应可由在两束泵浦场作用下探测场的透射谱进行表征。我们首先考虑光学腔被驱动至蓝边带而微波腔被驱动至红边带的情况,即 $\Delta_o = -\omega_m, \Delta_e = \omega_m$。图 4.2.2 所示是泵浦场功率 $P_o = 0, 10\ \mu\text{W}, 40\ \mu\text{W}$ 时探测场的透射谱随探测场-腔场失谐量 $\Delta_p = \Omega_p - \omega_o$ 变化的曲线。当泵浦场关闭时,从图 4.2.2(a)可以看出探测场透射谱呈现空腔时通常的洛伦兹线型。但是当 $P_o = 10\ \mu\text{W}$ 时,共振处探测场透射率比泵浦场关闭时的透射率还小,从图 4.2.2(b)的插图可以看得更清晰。这种现象称为光力诱导吸收,与原子系统中的电磁诱导吸收类似。当泵浦场功率更高时,如 $P_o = 40\ \mu\text{W}$,系统从光力诱导吸收转变为参量放大,此时探测场的透射率可以远大于1。在这种情况下,混杂系统可以用作晶体管[46]来放大弱的光或者微波信号。这些现象可以由相长干涉进行解释。泵浦场和探测场的共同作用产生了频率为拍频 $\delta = \Omega_p - \Omega_o$ 的辐射压力,并作用在共同的机械振子上。当这个调制频率与机

械振子的共振频率 ω_m 接近时,机械振子开始相干振动,并产生频率为 $\Omega_{as} = \Omega_o + \omega_m$ 的反斯托克斯场和频率为 $\Omega_s = \Omega_o - \omega_m$ 的斯托克斯场。如果光学腔被驱动至蓝边带($\Delta_o = \omega_o - \Omega_o = -\omega_m$),非共振的反斯托克斯散射被抑制而只存在斯托克斯散射。当入射的探测场与光学腔共振时,斯托克斯场和探测场之间的相长干涉增强了腔内的探测场。探测光子不断注入到腔内表现为透射率减小。当泵浦场功率足够强时,下转换的泵浦光子数远大于入射到腔内的探测光子数,混杂光力系统中探测场的参量放大可以导致探测场的透射率大于1。

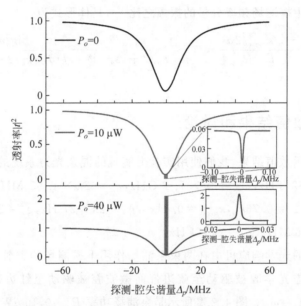

图 4.2.2 泵浦场功率 $P_o = 0, 10\ \mu\mathrm{W}, 40\ \mu\mathrm{W}$ 时探测场的透射谱 $|t|^2$ 随探测场-腔场失谐量 $\Delta_p = \Omega_p - \omega_o$ 变化的情况。光学腔被驱动至蓝边带而微波腔被驱动至红边带,即 $\Delta_o = -\omega_m, \Delta_e = \omega_m$。图 4.2.2(b) 和(c) 中的插图是共振区域附近探测场透射谱的放大图

(请扫 II 页二维码看彩图)

图 4.2.3 所示是峰值探测透射率 $|t|^2$ 随泵浦场功率变化的情况。从图中可以看出,当泵浦场功率从零开始增加时,峰值透射率逐渐减小到一个最小值,即出现了光力诱导吸收现象。但是当泵浦场功率进一步增加时,峰值透射率开始单调增加,并且可以出现透射率大于 1 的情况,即探测场被放大。因此,通过调节泵浦场的功率可以实现光力诱导吸收和参量放大之间的转换。

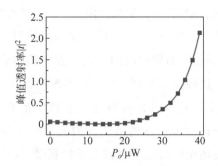

图 4.2.3　当 $\Delta_o=-\omega_m, \Delta_e=\omega_m$ 时腔共振处的峰值探测透射率 $|t|^2$ 随光学泵浦场功率 P_o 变化的情况。微波泵浦场功率 $P_e=1\,\mu W$，其他参数与图 4.2.2 相同

（请扫 II 页二维码看彩图）

该混杂光力系统对弱探测场的光学响应可以更加灵活地进行控制。接下来我们考虑光学腔被驱动至红边带而微波腔被驱动至蓝边带的情况（$\Delta_o=\omega_m$，$\Delta_e=-\omega_m$）。图 4.2.4 给出的是 $P_e=0,20\,nW,40\,nW,60\,nW$ 时探测场透射谱 $|t|^2$ 随探测失谐量 Δ_p 变化的情况。当微波泵浦场关闭时，微波腔中的辐射压对机械振子的作用可以忽略，则混杂光力系统变成典型的单腔光力系统。这里光学泵浦场功率取 $5\,\mu W$，可以看到共振区域探测场的透射率近似为 1。这种情况下系统中出现了光力诱导透明现象，并且透明窗口的宽度可以

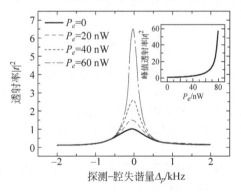

图 4.2.4　泵浦场功率 $P_e=0,20\,nW,40\,nW,60\,nW$ 时探测场透射谱 $|t|^2$ 随探测场-腔场失谐量 Δ_p 变化的情况。插图画的是探测场与光学腔共振时峰值透射率随泵浦场功率 P_e 变化的情况

（请扫 II 页二维码看彩图）

通过增加光学泵浦场功率进一步变宽。但是,如果微波泵浦场被打开,探测场的透射率可以大于1。比如,当 $P_e = 20$ nW 时,透射的探测场可以被放大150%。图4.2.4的插图表明探测场的峰值透射率随着微波泵浦场功率的增加而单调增加。出现这种现象的原因可以这样解释:当微波腔被驱动至蓝边带时,泵浦光子通过斯托克斯过程转换为声子和腔光子,进而导致机械振子的加热。产生的声子进一步被红边带驱动的光学腔中的反斯托克斯过程吸收。增加的微波泵浦场相干增强了机械振子的振动,导致了探测场的参量放大。基于以上的讨论,我们可以看到该混杂光力系统中的光学响应可以通过两束泵浦场的频率和功率进行有效的控制。

4.2.4 小结

本节研究了由一个光学腔和一个微波腔耦合于一个共同的机械振子形成的混杂光力系统中的光学响应[47]。当光学腔被驱动至蓝边带而微波腔被驱动至红边带时,斯托克斯场和探测场之间的相长干涉可以导致光力诱导吸收和放大现象,并且光力诱导吸收和放大之间可以通过调节光学泵浦场的功率进行转换。此外,当光学腔被驱动至红边带而微波腔被驱动至蓝边带时,微波泵浦场可以用作光力诱导透明和参量放大之间相互转换的开关。

参 考 文 献

[1] KIPPENBERG T J, VAHALA K J. Cavity optomechanics: back-action at the mesoscale[J]. Science, 2008, 321: 1172-1176.
[2] MARQUARDT F, GIRVIN S M. Optomechanics[J]. Physics, 2009, 2: 40.
[3] ASPELMEYER M, MEYSTRE P, SCHWAB K. Quantum optomechanics[J]. Phys. Today, 2012, 65: 29-35.
[4] LIU Y C, HU Y W, WONG C W, et al. Review of cavity optomechanical cooling[J]. Chin. Phys. B, 2013, 22: 114213.
[5] TEUFEL J D, DONNER T, LI D, et al. Sideband cooling of micromechanical motion to the quantum ground state[J]. Nature, 2011, 475: 359-363.
[6] CHAN J, ALEGRE T P, SAFAVI-NAEINI A H, et al. Laser cooling of a nanomechanical oscillator into its quantum ground state[J]. Nature, 2011, 478: 89-92.

[7] VERHAGEN E, DELGLISE S, WEIS S, et al. Quantum coherent coupling of a mechanical oscillator to an optical cavity mode[J]. Nature, 2012, 482: 63-67.

[8] PALOMAKI T A, HARLOW J W, TEUFEL J D, et al. Coherent state transfer between itinerant microwave fields and a mechanical oscillator[J]. Nature, 2013, 495: 210-214.

[9] DOBRINDT J M, WILSON-RAE I, KIPPENBERG T J. Parametric normal-mode splitting in cavity optomechanics[J]. Phys. Rev. Lett., 2008, 101: 263602.

[10] GRÖBLACHER S, HAMMERER K, VANNER M R, et al. Observation of strong coupling between a micromechanical resonator and an optical cavity field [J]. Nature, 2009, 460: 724-727.

[11] AGARWAL G S, HUANG S. Electromagnetically induced transparency in mechanical effects of light[J]. Phys. Rev. A, 2010, 81: 041803.

[12] WEIS S, RIVIÈRE R, DELÉGLISE S, et al. Optomechanically induced transparency [J]. Science, 2010, 330: 1520-1523.

[13] SAFAVI-NAEINI A H, ALEGRE T P M, CHAN J, et al. Electromagnetically induced transparency and slow light with optomechanics[J]. Nature, 2011, 472: 69-73.

[14] TEUFEL J D, LI D, ALLMAN M S, et al. Circuit cavity electromechanics in the strong-coupling regime[J]. Nature, 2011, 471: 204-208.

[15] SHU J. Electromagnetically induced transparency in an optomechanical system[J]. Chin. Phys. Lett., 2011, 28: 104203.

[16] YAN X B, GU K H, FU C B, et al. Electromagnetically induced transparency in a three-mode optomechanical system[J]. Chin. Phys. B, 2014, 23: 114201.

[17] FLEISCHHAUER M, IMAMOGLU A, MARANGOS J P. Electromagnetically induced transparency: optics in coherent media[J]. Rev. Mod. Phys., 2005, 77: 633-673.

[18] WU Y, YANG X X. Electromagnetically induced transparency in V-, Λ-, and cascade-type schemes beyond steady-state analysis [J]. Phys. Rev. A, 2005, 71: 053806.

[19] BOLLER K J, IMAMOGLU A, HARRIS S E. Observation of electromagnetically induced transparency[J]. Phys. Rev. Lett., 1991, 66: 2593-2596.

[20] PHILLIPS M C, WANG H, RUMYANTSEV I, et al. Electromagnetically induced transparency in semiconductors via biexciton coherence[J]. Phys. Rev. Lett., 2003, 91: 183602.

[21] SANTORI C, TAMARAT P, NEUMANN P, et al. Coherent population trapping of single spins in diamond under optical excitation[J]. Phys. Rev. Lett., 2006, 97: 247401.

[22] ZHOU X, HOCKE F, SCHLIESSER A, et al. Slowing, advancing and switching of microwave signals using circuit nanoelectromechanics[J]. Nat. Phys., 2013, 9: 179-

184.

[23] FIORE V, YANG Y, KUZYK M C, et al. Storing optical information as a mechanical excitation in a silica optomechanical resonator[J]. Phys. Rev. Lett., 2011,107: 133601.

[24] HOCKE F, ZHOU X, SCHLIESSER A, et al. Electromechanically induced absorption in a circuit nano-electromechanical system[J]. New J. Phys., 2012, 14: 123037.

[25] MASSEL F, HEIKKILÄ T T, PIRKKALAINEN J M, et al. Microwave amplification with nanomechanical resonators[J]. Nature,2011,480: 351-354.

[26] SINGH V, BOSMAN S J, SCHNEIDER B H, et al. Optomechanical coupling between a multilayer graphene mechanical resonator and a superconducting microwave cavity[J]. Nat. Nanotech.,2014,9: 820.

[27] TIAN L, WANG H L. Optical wavelength conversion of quantum states with optomechanics[J]. Phys. Rev. A,2010,82: 053806.

[28] TIAN L. Adiabatic state conversion and pulse transmission in optomechanical systems[J]. Phys. Rev. Lett.,2012,108: 153604.

[29] HILL J T, SAFAVI-NAEINI A H, CHAN J, et al. Coherent optical wavelength conversion via cavity optomechanics[J]. Nat. Commun.,2012,3: 1196.

[30] BARZANJEH S, ABDI M, MILBURN G J, et al. Reversible opticalto-microwave quantum interface[J]. Phys. Rev. Lett.,2012,109: 130503.

[31] MCGEE S A, MEISER D, REGAL C A, et al. Mechanical resonators for storage and transfer of electrical and optical quantum states [J]. Phys. Rev. A, 2013, 87: 053818.

[32] BOCHMANN J, VAINSENCHER A, AWSCHALOM D D, et al. Nanomechanical coupling between microwave and optical photons[J]. Nat. Phys.,2013,9: 712.

[33] ANDREWS R W, PETERSON R W, PURDY T P, et al. Reversible and efficient conversion between microwave and optical light[J]. Nat. Phys.,2013,10: 321.

[34] REGAL C A, LEHNERT K W. From cavity electromechanics to cavity optomechanics[J]. J. Phys. Conf. Ser.,2011,264: 012025.

[35] LU X Y, ZHANG W M, ASHHAB S, et al. Quantum-criticality-induced strong Kerr nonlinearities in optomechanical systems[J]. Sci. Rep.,2013,3: 2943.

[36] LI H K, REN X X, LIU Y C, et al. Photon-photon interactions in a largely detuned optomechanical cavity[J]. Phys. Rev. A,2013,88: 053850.

[37] QU K N, AGARWAL G S. Phonon-mediated electromagnetically induced absorption in hybrid opto-electromechanical systems[J]. Phys. Rev. A,2013,87: 031802(R).

[38] NUNNENKAMP A, SUDHIR V, FEOFANOV A K, et al. Quantum limited amplification and parametric instability in the reversed dissipation regime of cavity optomechanics[J]. Phys. Rev. Lett.,2014,113: 023604.

[39] LEZAMA A, BARREIRO S, AKULSHIN A M. Electromagnetically induced absorption[J]. Phys. Rev. A,1999,59: 4732.

[40] METELMANN A, CLERK A A. Quantum-limited amplification via reservoir engineering[J]. Phys. Rev. Lett. ,2014,112: 133904.

[41] MA J Y, YOU C, SI L G, et al. Optomechanically induced transparency in the mechanical-mode splitting regime[J]. Opt. Lett. ,2014,39: 4180.

[42] GENES C, VITALI D, TOMBESI P, et al. Ground-state cooling of a micromechanical oscillator: comparing cold damping and cavity-assisted cooling schemes[J]. Phys. Rev. A,2008,77: 033804.

[43] XIONG H,SI L G,ZHENG A S,et al. Higher-order sidebands in optomechanically induced transparency[J]. Phys. Rev. A,2012,86: 013815.

[44] BOYD R W. Nonlinear optics[M]. San Diego: Academic,2008.

[45] GARDINER C W,ZOLLER P. Quantum noise[M]. New York: Springer,2004.

[46] CHEN B, JIANG C, LI J J, et al. All-optical transistor based on a cavity optomechanical system with a Bose-Einstein condensate[J]. Phys. Rev. A,2011, 84: 055802.

[47] JIANG C,CUI Y S,LIU H X,et al. Controllable optical response in hybrid opto-electromechanical systems[J]. Chin. Phys. B,2016,24: 054206.

第 5 章 基于混杂光-电力系统的超灵敏纳米机械质量传感器

5.1 引　言

纳米机械振子由于其极小的质量($10^{-15} \sim 10^{-21}$ kg)、较高的共振频率(kHz～GHz)以及超高的品质因素($10^3 \sim 10^7$)[1,2]被广泛应用于灵敏的生物和化学传感器。纳米机械振子质量传感器主要利用追踪吸附粒子后质量变化引起的纳米机械振子的共振频率移动。质量灵敏度是一个将频率移动和增加的质量联系起来的非常重要的参数,它正比于振子的共振频率,反比于它的质量[3]。这促使质量分辨率从最开始的 fg(10^{-15} g)[4]到后来的 ag(10^{-18} g)[5]、zg(10^{-21} g)[6]、到最近的 yg(10^{-24} g)[7],被不断地提高。在过去十几年中,基于悬臂[8,9]、纳米线[10-12]、悬挂微通道振子[13,14]以及碳纳米管[15,16]的纳米机械振子质量传感器已被广泛用于测量单分子、病毒以及纳米颗粒的质量。最近,刘等用一个支持回音壁模式的微环形光机械振子实现了亚皮克量级的质量测量[17,18],他们的分析表明飞克水平的分辨率是可以实现的。此外,邵(Shao)等在实验上利用追踪环形微腔中的回音壁(WGM)模式展宽探测出了单个纳米颗粒以及慢病毒,而这一过程不受来自探测光以及环境扰动带来的噪声[19]。李和朱最近在理论上提出了一种基于环形微腔光机械系统的非线性质量传感器方案[20,21]。

腔光力系统主要由机械振子与电磁腔通过辐射压力耦合形成,近年来受到了广泛的研究[22-24]。由于机械振子引起的相互作用改变了光机系统的光学响应特性,导致了正则模式分裂[25]和电磁感应透明(EIT)现象[26-28]。同样,光力相互作用可以用来灵敏地读取机械运动[29,30],机械振子的量子基态冷却[31,32],以及光和机械振子之间的量子相干耦合[33,34]。最近,安德鲁斯(Andrews)等成功地将一个微机械振子同时耦合于一个微波腔和一个光学腔[35]，从而构成了一个混杂腔光-电力系统[36-38]。振子的机械振动改变光学腔和微波腔的共振频率。当两个腔同时受到适当频率和功率的泵浦场驱动时，一个弱的探测场被用来扫描光学共振频率，从而机械振子的共振频率可以通过探测场的透射谱得到。因此，吸附在振子上的粒子的质量可以通过测量频率移动得到。机械振子非常窄的线宽使得它可以被用作高分辨率的质量传感器。与刘等的质量测量类似[17,18]，这里提出的质量传感器也是基于测量吸附粒子前后振子共振频率的移动。但是在刘等的实验中只使用了一束激光来激发和灵敏地追踪机械振子的共振频率。而在我们这里所考虑的混杂光-电力系统中，两束强的泵浦场被用来驱动腔场，而一束弱的探测场被用来测量机械振子共振频率的移动。这样的泵浦-探测技术已经被广泛应用于光力实验中[26-28,35]。

5.2 模型和理论

考虑的混杂光-电力系统由一个微波腔和一个光学腔同时耦合于一个共同的机械振子形成。系统的示意图如图 5.2.1 所示，其中振子由一个可以自由振动的硅氮薄膜形成，共振频率为 ω_m，衰减率为 γ_m。共振频率为 ω_o 的光学腔同时受到振幅为 E_o、频率为 Ω_o 的强的泵浦场和振幅为 E_p、频率为 Ω_p 的弱的探测场驱动。由等效电感 L 和等效电容 C 表示的共振频率为 ω_e 的微波腔只受到一束振幅为 E_e、频率为 Ω_e 的强的泵浦场驱动。当薄膜受到辐射压力的作用沿着腔内驻波场运动时，光学腔的共振频率将发生变化[39]。同时，振动的薄膜改变微波腔的电容，从而进一步改变微波腔的共振频率。$g_o(g_e)$ 是力学模式和光（微波）腔之间的单光子耦合强度。在泵浦场频率 Ω_o 和 Ω_e 的旋转框架下，混杂系统的哈密顿量可表示为

$$H = H_0 + H_{\text{int}} + H_{\text{drive}}$$

式中，

$$\begin{aligned}
H_0 =& \hbar \Delta_o a^+ a + \hbar \Delta_e b^+ b + \hbar \omega_m c^+ c \\
H_{\text{drive}} =& i\hbar\sqrt{\kappa_{o,\text{ext}}} E_o (a^+ - a) + i\hbar\sqrt{\kappa_{e,\text{ext}}}(b^+ - b) + \\
& i\hbar\sqrt{\kappa_{o,\text{ext}}} E_p (a^+ e^{-i\delta t} - a e^{i\delta t})
\end{aligned} \quad (5.2.1)$$

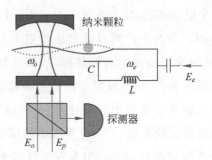

图 5.2.1 混杂光-电力系统示意图。一个机械振子共同耦合于一个光学腔和一个微波腔。光学腔同时受到强的泵浦场和弱的探测场作用,而微波腔只受到一束强的泵浦场作用
(请扫Ⅱ页二维码看彩图)

这里,$\Delta_o = \omega_o - \Omega_o$,$\Delta_e = \omega_e - \Omega_e$ 分别是相应的腔-泵浦场失谐量。H_{drive} 表示输入场与腔场之间的相互作用强度,其中 $\delta = \Omega_p - \Omega_o$ 是探测激光束与泵浦激光束之间的失谐量。E_o、E_e 和 E_p 是输入场的振幅,它们与各自的功率之间的关系为 $|E_o| = \sqrt{2P_o\kappa_o/\hbar\Omega_o}$,$|E_e| = \sqrt{2P_e\kappa_e/\hbar\Omega_e}$,$|E_p| = \sqrt{2P_p\kappa_o/\hbar\Omega_p}$,其中 $\kappa_o(\kappa_e)$ 是光(微波)腔模式的线宽。$\kappa_{o,\text{ext}}(\kappa_{e,\text{ext}})$ 描述的是光(微波)腔中的能量传输至传播场中的速率[35]。描述系统的量子朗之万方程可以通过求解海森伯运动方程并加上腔场和力学模式相应的衰减和噪声项得到,如下所示:

$$\dot{a} = -\mathrm{i}(\Delta_o - g_0 Q)a - \kappa_o a + \sqrt{\kappa_{o,\text{ext}}}(E_o + E_p \mathrm{e}^{-\mathrm{i}\delta t}) + \sqrt{2\kappa_o}a_{in} \quad (5.2.2)$$

$$\dot{b} = -\mathrm{i}(\Delta_e - g_e Q)b - \kappa_e b + \sqrt{\kappa_{e,\text{ext}}}E_e + \sqrt{2\kappa_e}b_{in} \quad (5.2.3)$$

$$\ddot{Q} + \gamma_m \dot{Q} + \omega_m^2 Q = 2g_o \omega_m a^+ a + 2g_e \omega_m b^+ b + \xi \quad (5.2.4)$$

式中,$Q = c^+ + c$,a_{in} 和 b_{in} 分别是平均值为零的输入真空噪声,在时域空间遵循以下的关联函数:

$$\langle a_{in}(t)a_{in}^+(t')\rangle = \langle b_{in}(t)b_{in}^+(t')\rangle = \delta(t-t') \quad (5.2.5)$$

$$\langle a_{in}^+(t)a_{in}(t')\rangle = \langle b_{in}^+(t)b_{in}(t')\rangle = 0 \quad (5.2.6)$$

力学模式受到黏滞力和平均值为零的满足如下关联函数的布朗随机力作用[40]:

$$\langle \xi(t)\xi(t')\rangle = \frac{\gamma_m}{\omega_m}\int \frac{\mathrm{d}\omega}{2\pi}\omega \mathrm{e}^{-\mathrm{i}\omega(t-t')}\left[1 + \coth\left(\frac{\hbar\omega}{2k_\mathrm{B}T}\right)\right] \quad (5.2.7)$$

其中,k_B 是玻耳兹曼常数,T 是机械振子库的温度。我们通过令方程(5.2.2)~

第 5 章　基于混杂光-电力系统的超灵敏纳米机械质量传感器

方程(5.2.4)的时间求导项为零得到如下稳态解：

$$a_s = \frac{\sqrt{\kappa_{o,\text{ext}}}E_o}{\kappa_o + i\Delta'_o}, \quad b_s = \frac{\sqrt{\kappa_{e,\text{ext}}}E_e}{\kappa_e + i\Delta'_e}, \quad Q_s = \frac{2}{\omega_m}(g_o|a_s|^2 + g_e|a_s|^2)$$

(5.2.8)

其中，$\Delta'_o = \Delta_o - g_0 Q_s$，$\Delta'_e = \Delta_e - g_e Q_s$ 是包含了辐射压效应后的有效的腔失谐量。接下来我们采用标准的方法来微扰求解方程(5.2.2)～方程(5.2.4)，将每个海森伯算符写成其稳态平均值与一个平均值为零的小的扰动项的和的形式，即

$$a = a_s + \delta a, \quad b = b_s + \delta b, \quad Q = Q_s + \delta Q \quad (5.2.9)$$

将上式代入量子朗之万方程(5.2.2)～方程(5.2.4)，并假设$|a_s| \gg 1, |b_s| \gg 1$，可以得到以下的线性化的海森伯-朗之万方程：

$$\delta\dot{a} = -(\kappa_o + i\Delta_o)\delta a + ig_o Q_s \delta a + \sqrt{\kappa_{o,\text{ext}}}E_p e^{-i\delta t} + \sqrt{2\kappa_o}a_{in} \quad (5.2.10)$$

$$\delta\dot{b} = -(\kappa_e + i\Delta_e)\delta b + ig_e Q_s \delta a + ig_e b_s \delta Q + \sqrt{2\kappa_e}b_{in} \quad (5.2.11)$$

$$\delta\ddot{Q} + \gamma_m \delta\dot{Q} + \omega_m^2 \delta Q = 2g_o \omega_m a_s(\delta a + \delta a^+) + 2g_e \omega_m b_s(\delta b^+ + \delta b) + \xi$$

(5.2.12)

其中我们忽略了 $\delta a^+ \delta a$、$\delta b^+ \delta b$、$\delta a \delta Q$ 和 $\delta b \delta Q$ 等非线性项，这些项可以导致二阶和高阶边带等有趣的现象[41]。接下来，因为驱动场是经典的相干场，我们将把所有的算符用它们的期待值代替并且舍弃量子和热噪声项[26]，这样线性化的朗之万方程可改写为

$$\langle\delta\dot{a}\rangle = -(\kappa_o + i\Delta_o)\langle\delta a\rangle + ig_o Q_s\langle\delta a\rangle + ig_o a_s\langle\delta Q\rangle + \sqrt{\kappa_{o,\text{ext}}}E_p e^{-i\delta t}$$

(5.2.13)

$$\langle\delta\dot{b}\rangle = -(\kappa_e + i\Delta_e)\langle\delta b\rangle + ig_e Q_s\langle\delta b\rangle + ig_e b_s\langle\delta Q\rangle \quad (5.2.14)$$

$$\langle\delta\ddot{Q}\rangle + \gamma_m\langle\delta\dot{Q}\rangle + \omega_m^2\langle\delta Q\rangle = 2g_o\omega_m a_s(\langle\delta a\rangle + \langle\delta a^+\rangle) + 2g_e\omega_m b_s(\langle\delta b^+\rangle + \langle\delta b\rangle) \quad (5.2.15)$$

为了求解方程(5.2.13)～方程(5.2.15)，作以下代换[42] $\langle\delta a\rangle = a_+ e^{-i\delta t} + a_- e^{i\delta t}$，$\langle\delta b\rangle = b_+ e^{-i\delta t} + b_- e^{i\delta t}$，并且 $\langle\delta Q\rangle = Q_+ e^{-i\delta t} + Q_- e^{i\delta t}$。将以上代换代入方程(5.2.13)～方程(5.2.15)，可以得到以下形式的解：

$$a_+ = \frac{\sqrt{\kappa_{o,\text{ext}}}E_p}{\kappa_o + i\Delta'_o - i\delta} - \frac{ig_o^2 n_o}{f(\delta)} \frac{\sqrt{\kappa_{o,\text{ext}}}E_p}{(\kappa_o + i\Delta'_o - i\delta)^2} \quad (5.2.16)$$

$$a_- = -\frac{1}{f(\delta)^*(\kappa_o + i\Delta'_o)^2} \cdot \frac{ig_o^2 \kappa_{o,\text{ext}} E_o^2 \sqrt{\kappa_{o,\text{ext}}} E_p}{(\kappa_o + i\delta)^2 + \Delta'^2_o} \quad (5.2.17)$$

式中,

$$f(\delta) = \frac{2\Delta'_o g_o^2 n_o}{(\kappa_o + i\delta)^2 + \Delta'^2_o} + \frac{2\Delta'_e g_e^2 n_e}{(\kappa_e - i\delta)^2 + \Delta'^2_e} - \frac{\omega_m^2 - \delta^2 - i\delta\gamma_m}{\omega_m}$$

$$(5.2.18)$$

这里,$n_o = |a_s|^2, n_e = |b_s|^2$ 分别表示每个腔内的泵浦光子数,由以下耦合方程决定:

$$n_o = \frac{\kappa_{o,\text{ext}} E_o^2}{\kappa_o^2 + [\Delta_o - 2g_o/\omega_m(g_o n_o + g_e n_e)]^2} \quad (5.2.19)$$

$$n_e = \frac{\kappa_{e,\text{ext}} E_e^2}{\kappa_e^2 + [\Delta_e - 2g_e/\omega_m(g_o n_o + g_e n_e)]^2} \quad (5.2.20)$$

从腔中出来的输出场可以由标准的输入-输出理论得到[43],即 $a_{\text{out}}(t) = a_{\text{in}}(t) - \sqrt{\kappa_{ex}} a(t)$,其中 $a_{\text{out}}(t)$ 是输出场算符。考虑光学腔的输出场,有

$$\langle a_{\text{out}}(t)\rangle = (E_o - \sqrt{\kappa_{o,\text{ext}}} a_s)e^{-i\Omega_o t} + (E_p - \sqrt{\kappa_{o,\text{ext}}} a_+)e^{-i(\delta+\Omega_o)t} - \sqrt{\kappa_{o,\text{ext}}} a_- e^{i(\delta+\Omega_o)t}$$
$$= (E_o - \sqrt{\kappa_{o,\text{ext}}} a_s)e^{-i\Omega_o t} + (E_p - \sqrt{\kappa_{o,\text{ext}}} a_+)e^{-i\Omega_p t} - \sqrt{\kappa_{o,\text{ext}}} a_- e^{-i(2\Omega_o - \Omega_p)t}$$

$$(5.2.21)$$

从方程(5.2.21)可以看出,输出场中包含两个输入分量(Ω_o 和 Ω_p)和一个新产生的频率为 $2\Omega_o - \Omega_p$ 的四波混频(FWM)分量。四波混频是一种三阶非线性过程[42],其中两个频率为 Ω_o 的泵浦光子通过光子与声子之间的光机械相互作用转变成一个频率为 Ω_{idler} 的闲置光子和一个频率为 Ω_p 的探测光子:$\Omega_{\text{idler}} = 2\Omega_o - \Omega_p$。这一过程已经在超高品质因素的环形微腔实验中被观察到[44],并在单模光机械系统中被从理论上研究[45]。基于现有的实验条件,四波混频应该可以在混杂光-电机械系统中被观察到。探测场的透射率定义为输出场中探测场频率分量与输入探测场之间的比值,由下式给出:

$$t(\Omega_p) = \frac{E_p - \sqrt{\kappa_{o,\text{ext}}} a_+}{E_p} = 1 - \left[\frac{\kappa_{o,\text{ext}}}{\kappa_o + i\Delta'_o - i\delta} - \frac{1}{f(\delta)}\frac{ig_o^2 n_o \kappa_{o,\text{ext}}}{(\kappa_o + i\Delta'_o - i\delta)^2}\right]$$

$$(5.2.22)$$

同样,根据探测场定义的四波混频强度由下式给出:

$$\text{FWM} = \left| \frac{\sqrt{\kappa_{o,\text{ext}}} a_-}{E_p} \right|^2 = \left| -\frac{1}{f(\delta)^* (\kappa_o + i\Delta_o')^2} \cdot \frac{ig_o^2 \kappa_{o,\text{ext}}^2 E_o^2}{(\kappa_o + i\delta)^2 + \Delta_o'^2} \right|^2$$

(5.2.23)

接下来,我们将提出一个基于测量探测场透射谱和四波混频谱从而得出吸附物质量的方案,其中薄膜的质量可以从谱线中得出。吸附在薄膜表面的粒子将会导致共振频率的移动,频率移动量与吸附质量之间的关系为[3]

$$\Delta m = -\frac{2m_{\text{eff}}}{\omega_m} \Delta \omega = R^{-1} \Delta \omega \tag{5.2.24}$$

其中,$R = (-2m_{\text{eff}}/\omega_m)^{-1}$ 定义为质量响应率。

5.3 结果和讨论

为了说明数值结果,我们选择实验上可以实现的混杂光-电力系统。模拟中所用的参数为[31,35]:$\omega_o = 2\pi \times 282$ THz, $\omega_e = 2\pi \times 7.1$ GHz, $\kappa_o = 2\pi \times 1.65$ MHz, $\kappa_e = 2\pi \times 1.6$ MHz, $\kappa_{o,\text{ext}} = 0.76\kappa_o$, $\kappa_{e,\text{ext}} = 0.11\kappa_e$, $g_o = 2\pi \times 27$ Hz, $g_e = 2\pi \times 2.7$ Hz, $\omega_m = 2\pi \times 5.6$ MHz, 振子的有效质量约为 $m_{\text{eff}} = 45$ pg。我们以杆状病毒和金纳米颗粒(金的密度是 $\rho_{\text{Au}} = 19\ 300$ kg/m^3)作为吸附样品。需要注意的是,该混杂系统必须工作在低温环境下以保持超导电路工作良好。

质量测量主要是基于探测增加的吸附物吸附前后机械振子共振频率的移动量。首先,我们提出一种基于混杂光-电力系统测量机械振子共振频率的方法。图 5.3.1(a)给出了当 $\Delta_o = \Delta_e = 0$, $P_o = P_e = 0.01$ μW 时探测场透射谱随探测-腔场失谐量 Δ_p 变化的情况。从该图可以清晰地看到当探测场与光学腔共振时出现一个宽的透射凹陷,这对应于腔场的吸收。此外,$\Delta_p = \pm \omega_m$ 时有两个尖锐的边带尖峰,它们代表力学模式的共振放大和吸收。边带峰的谱线宽度是力学模式的衰减率 $\gamma_m/(2\pi) = 4$ Hz,这可以从图 5.3.1(b)中的放大图清晰地看到。这种非常窄的谱线宽度非常有利于分辨由于吸附质量引起的频率移动。因此,图 5.3.1 提供了一种测量机械振子共振频率的有效方法。测量过程描述如下:①对光学腔和微波腔分别施加一束强的光学泵浦场和一束强的微波泵浦场,并且固定泵浦场的频率与腔场频率一致,即($\Delta_o = \Delta_e = 0$);

② 对光学腔施加另外一束弱的探测场,在腔场频率附近扫描探测场频率。通过观察探测场透射谱,我们可以很容易得到机械振子的共振频率。这一现象的物理机制可以理解为探测场-泵浦场失谐量 δ 为振动频率的辐射压力作用的结果。如果 δ 接近于机械振子共振频率 ω_m,机械振子开始相干振动起来。相干振动将导致泵浦场产生两个机械边带,进一步与探测场之间产生干涉,从而修正探测场透射谱。

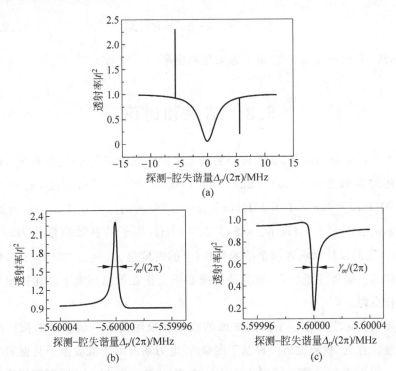

图 5.3.1 (a)探测场透射谱随探测场-腔场失谐量 Δ_p 变化的曲线;
(b),(c)图(a)中两个边带峰的放大图

从探测场透射谱中得出机械振子的共振频率之后,接下来测量由于物体落到振子上之后产生的频率移动。频率的移动量取决于吸附物体的质量以及吸附的位置,所以将一个物体吸附到一个已知的位置上之后就可以直接得出它的质量。为了简单起见,我们假设吸附物体的质量与频率移动量之间满足方程(5.2.24)[7]。如果一个杆状病毒落到机械振子的表面上,振子的总质量将会增加,相应的共振频率将会变小。图 5.3.2 所示是吸附单个杆状病毒

吸附到机械振子表面前后共振频率附近探测场透射谱随探测场-腔场失谐量 Δ_p 变化的情况。可以看出由于机械振子的质量变大,透射谱中出现了 $\Delta\omega = -2\pi \times 93$ Hz 的共振频率移动。基于这个频率移动量并根据方程(5.2.24),我们发现增加的质量为 1.5 fg(1 fg $= 10^{-15}$ g),大概为单个杆状病毒的质量[46]。因此,这里研究的混杂光-电机械系统可用于测量单个病毒的质量。尽管我们这里提出的质量测量的方案是简单、可行的,但在实验中想要实现这样的质量传感器还存在诸多困难。比如,杆状病毒和机械振子在吸附之前需经过特定过程的处理[46]。此外,质量灵敏度 R 是评价机械振子用于质量测量时非常重要的一个参数。图 5.3.2 插图显示共振频率移动量与吸附到机械振子上病毒数量之间存在线性关系,这样的线性关系已经在多种实验中被证实[6,8]。线的负的斜率给出了振子的质量灵敏度。小的质量和高的共振频率在获得较高的质量灵敏度中起着至关重要的作用。孙和郑等最近设计并实验上通过将一个质量为 25 fg 的两边夹住的纳米机械双臂振子嵌入到一个精细调制的二维光子晶体中,演示了一个飞克 L3 纳米横梁型光机械腔,其中光学腔的基模频率约为 1 GHz[47-49]。他们也指出这样的飞克质量、高机械共振频率结构可以被用作超灵敏的质量、力和位移传感器。

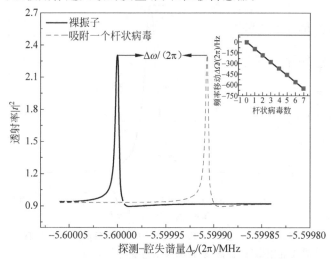

图 5.3.2 单个杆状病毒吸附到机械振子表面前后探测场透射谱随探测场-腔场失谐量变化的曲线。插图展示的是频率移动量与吸附病毒数量之间的关系

(请扫 II 页二维码看彩图)

此外,基于三阶非线性光学效应,混杂光-电机械系统可以被用作一种非线性质量传感器。图 5.3.3 所示是振子上吸附一个直径为 20 nm 的金纳米颗粒前后四波混频谱随着探测场-腔场失谐量变化的情况,其中 $\Delta_o=\Delta_e=0$。类似地,由于量子干涉效应,一个窄的峰出现在 $\Delta_p=\omega_m$ 处,这同样可用于测量机械振子的共振频率。在吸附了单个纳米颗粒之后,有一个 $\Delta\omega=-2\pi\times 5$ Hz 的频率移动,并且虚线的尖峰位于 $\omega_m+\Delta\omega$。根据方程(5.2.24),可以得到单个 20 nm 直径的金纳米颗粒的质量:$\Delta m=-\dfrac{2m_{\text{eff}}}{\omega_m}\Delta\omega=80.4$ ag(1 ag= 10^{-18} g),这一数值接近于根据密度和体积计算出的金纳米颗粒的质量。与传统质谱仪相比,基于四波混频效应的非线性质量传感器有一些明显的优势。第一,粒子不需要像传统质谱仪那样首先得离子化[50],并且它们的质量可以从四波混频谱中很容易测出来。第二,泵浦场和探测场同时作用产生了拍频波去驱动机械振子。因此,不管是高频还是低频的振子都适用于这里提出的质量测量方案。第三,有探测噪声存在时使用非线性光谱可能比线性光谱更具优势[51]。

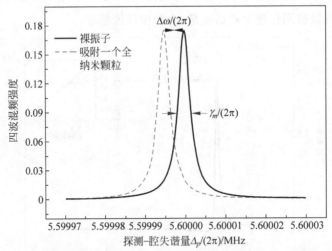

图 5.3.3　振子上吸附单个直径为 20 nm 的金纳米颗粒前后非线性探测场透射谱
(四波混频谱)随探测场-腔场失谐量变化的曲线
(请扫Ⅱ页二维码看彩图)

最后,需要指出的是,我们在数值模拟过程中并没有考虑噪声过程的影响。实际上,各种噪声源会对机械振子的最终质量灵敏度产生影响,比如由于振子的内部损耗机制引起的热机械噪声,残留分子引起的吸附-解吸附噪声[3,52],以及读取电路中的探测噪声[51]。在本节研究的混杂电机械系统中,机械运动的热噪声是最主要的噪声源。如果相关实验可以在 40 mK 的极低温下进行,热噪声将被大大消除。此外,增加的振动噪声可以通过使用高品质因素的机械振子而减少[35]。

5.4 小　　结

本节理论上证实了由一个光学腔和一个微波腔耦合于一个共同的机械振子组成的混杂光-电力系统可以被用作超灵敏的质量传感器[53]。由于力学模式与两个光场的拍频之间的量子干涉效应,机械振子的共振频率可以通过探测场的透射谱和四波混频谱得到。因此,吸附到振子上的单个病毒或者纳米颗粒的质量可以通过增加的质量与相应的频率移动之间的关系得到。本节提出的方案在现有的实验条件下可以实现。

参 考 文 献

[1] SCHWAB K C, ROUKES M L. Putting mechanics into quantum mechanics[J]. Physics Today, 2005, 58: 36-42.

[2] ARLETT J L, MYERS E B, ROUKES M L. Comparative advantages of mechanical biosensors[J]. Nat. Nanotechnol. , 2011, 6: 203-215.

[3] EKINCI K L, TANG Y T, ROUKES M L. Ultimate limits to inertial mass sensing based upon nanoelectromechanical systems [J]. J. Appl. Phys. , 2004, 95: 2682-2689.

[4] LAVRIK N V, DATSKOS P G. Femtogram mass detection using photothermally actuated nanomechanical resonators[J]. Appl. Phys. Lett. , 2003, 82: 2697-2699.

[5] ILIC B, CRAIGHEAD H G, KRYLOV S, et al. Attogram detection using nanoelectromechanical oscillators[J]. J. Appl. Phys. , 2004, 95: 3694-3703.

[6] YANG Y T, CALLEGARI C, FENG X L, et al. Zeptogram-scale nanomechanical mass sensing[J]. Nano Lett. , 2006, 6: 583-586.

[7] CHASTE J, EICHLER A, MOSER J, et al. A nanomechanical mass sensor with yoctogram resolution[J]. Nat. Nanotechnol. , 2012, 7: 301-304.

[8] GUPTA A, AKIN D, BASHIR R. Single virus particle mass detection using

microresonators with nanoscale thickness[J]. Appl. Phys. Lett. ,2004,84: 1976-1978.

[9] LI M,TANG H X,ROUKES M L. Ultra-sensitive NEMS-based cantilevers for sensing,scanned probe and very high-frequency applications[J]. Nat. Nanotechnol. ,2007,2: 114-120.

[10] FENG X L,HE R,YANG P D,et al. Very high frequency silicon nanowire electromechanical resonators[J]. Nano Lett. ,2007,7: 1953-1959.

[11] NAIK A K,HANAY M S,HIEBERT W K,et al. Towards single-molecule nanomechanical mass spectrometry[J]. Nat. Nanotechnol. ,2009,4: 445-450.

[12] GIL-SANTOS E,RAMOS D,MARTINEZ J,et al. Nanomechanical mass sensing and stiffness spectrometry based on two-dimensional vibrations of resonant nanowires[J]. Nat. Nanotechnol. ,2010,5: 641-645.

[13] BURG T P,GODIN M,KNUDSEN S M,et al. Weighing of biomolecules,single cells and single nanoparticles in fluid[J]. Nature,2007,446: 1066-1069.

[14] OLCUMA S,CERMAK N,WASSERMAN S C,et al. Weighing nanoparticles in solution at the attogram scale[J]. Proc. Natl. Acd. Sci. ,2014,111: 1310-1315.

[15] LASSAGNE B,GARCIA-SANCHEZ D,AGUASCA A,et al. Ultrasensitive mass sensing with a nanotube electromechanical resonator[J]. Nano Lett. ,2008,8: 3735-3738.

[16] JENSEN K,KIM K,ZETTL A. An atomic-resolution nanomechanical mass sensor [J]. Nat. Nanotechnol. ,2008,3: 533-537.

[17] LIU F,HOSSEIN-ZADEH M. Mass sensing with optomechanical oscillation[J]. IEEE Sensors,2013,13: 146-147.

[18] LIU F,ALAIE S,LESEMAN Z C,et al. Sub-pg mass sensing and measurement with an optomechanical oscillator[J]. Opt. Express,2013,21: 19555-19567.

[19] SHAO L,JIANG X F,YU X C,et al. Detection of single nanoparticles and lentiviruses using microcavity resonance broadening[J]. Adv. Mater. ,2013,25 (39): 5616-5620.

[20] LI J J,ZHU K D. Nonlinear optical mass sensor with an optomechanical microresonator[J]. Appl. Phys. Lett. ,2012,101: 141905.

[21] LI J J,ZHU K D. All-optical mass sensing with coupled mechanical resonator systems[J]. Phys. Rep. ,2013,525: 223-254.

[22] KIPPENBERG T J,VAHALA K J. Cavity optomechanics: back-action at the mesoscale[J]. Science,2008,321: 1172-1176.

[23] MARQUARDT F,GIRVIN S M. Optomechanics[J]. Physics,2009,2: 40.

[24] ASPELMEYER M,MEYSTRE P,SCHWAB K. Quantum optomechanics[J]. Phys. Today,2012,65: 29-35.

[25] GROBLACHER S,HAMMERER K,VANNER M R,et al. Observation of strong coupling between a micromechanical resonator and an optical cavity field[J].

Nature,2009,460: 724-727.

[26] WEIS S,RIVIERE R,DELEGLISE S,et al. Optomechanically induced transparency [J]. Science,2010,330: 1520-1523.

[27] SAFAVI-NAEINI A H, ALEGRE T P M, CHAN J, et al. Electromagnetically induced transparency and slow light with optomechanics[J]. Nature,2011,472: 69-73.

[28] TEUFEL J D, LI D, ALLMAN M S, et al. Circuit cavity electromechanics in the strong-coupling regime[J]. Nature,2011,471: 204-208.

[29] ARCIZET O, COHADON P F, BRIANT T, et al. High-sensitivity optical monitoring of a micromechanical resonator with a quantum-limited optomechanical sensor[J]. Phys. Rev. Lett. ,2006,97: 133601.

[30] TEUFEL J D, DONNER T, CASTELLANOS-BELTRAN M A, et al. Nanomechanical motion measured with an imprecision below that at the standard quantum limit[J]. Nat. Nanotechnol. ,2009,4: 820-823.

[31] TEUFEL J D,DONNER T,LI D,et al. Sideband cooling of micromechanical motion to the quantum ground state[J]. Nature,2011,475: 359-363.

[32] CHAN J, ALEGRE T P, SAFAVI-NAEINI A H, et al. Laser cooling of a nanomechanical oscillator into its quantum ground state[J]. Nature, 2011, 478: 89-92.

[33] VERHAGEN E, DELEGLISE S, WEIS S, et al. Quantum-coherent coupling of a mechanical oscillator to an optical cavity mode[J]. Nature,2012,482: 63-67.

[34] PALOMAKI T A, HARLOW J W, TEUFEL J D, et al. Coherent state transfer between itinerant microwave fields and a mechanical oscillator[J]. Nature, 2013, 495: 210-214.

[35] ANDREWS R W,PETERSON R W,PURDY T P,et al. Bidirectional and efficient conversion between microwave and optical light[J]. Nat. Phys. ,2014,10: 321-326.

[36] REGAL C A, LEHNERT K W. From cavity electromechanics to cavity optomechanics[J]. J. Phys. Conf. Ser. ,2011,264: 012025.

[37] LU X Y, ZHANG W M, ASHHAB S, et al. Quantum-criticality-induced strong Kerr-nonlinearities in optomechanical systems[J]. Sci. Rep. ,2013,3: 2943.

[38] QU K N, AGARWAL G S. Phonon-mediated electromagnetically induced absorption in hybrid opto-electromechanical systems[J]. Phys. Rev. A,2013,87: 031802(R).

[39] THOMPSON J D,ZWICKL B M,JAYICH A M,et al. Strong dispersive coupling of a high-finesse cavity to a micromechanical membrane[J]. Nature, 2008, 45: 72-75.

[40] GENES C, VITALI D, TOMBESI P, et al. Ground-state cooling of a micromechanical oscillator: comparing cold damping and cavity-assisted cooling schemes[J]. Phys. Rev. A,2008,77: 033804.

[41] XIONG H, SI L G, ZHENG A S, et al. Higher-order sidebands in optomechanically induced transparency[J]. Phys. Rev. A, 2012, 86: 013815.

[42] BOYD R W. Nonlinear optics[M]. San Diego: Academic, 2008.

[43] GARDINER C W, ZOLLER P. Quantum noise[M]. New York: Springer, 2004.

[44] KIPPENBERG T J, SPILLANE S M, VAHALA K J. Kerr-nonlinearity optical parametric oscillation in an ultrahigh-Q toroid microcavity[J]. Phys. Rev. Lett., 2004, 93: 083904.

[45] HUANG S, AGARWAL G S. Normal-mode splitting and antibunching in Stokes and anti-Stokes processes in cavity optomechanics: radiation-pressure-induced four-wave-mixing cavity optomechanics[J]. Phys. Rev. A, 2010, 81: 033830.

[46] ILIC B, YANG Y, CRAIGHEAD H G. Virus detection using nanoelectromechanical devices[J]. Appl. Phys. Lett., 2004, 85: 2604-2606.

[47] SUN X, ZHENG J, POOT M, et al. Femtogram doubly clamped nanomechanical resonators embedded in a high-Q two-dimensional photonic crystal nanocavity[J]. Nano Lett., 2012, 12: 2299.

[48] ZHENG J, SUN X, POOT M, et al. Dispersive coupling and optimization of femtogram L3-nanobeam optomechanical cavities[C]. Frontiers in Optics, 2012.

[49] ZHENG J, SUN X, LI Y, et al. Femtogram dispersive L_3-nanobeam optomechanical cavities: design and experimental comparison[J]. Opt. Express, 2012, 20: 26486-26498.

[50] BOISEN A. Nanoelectromechanical systems: mass spec goes nanomechanical[J]. Nat. Nanotechnol., 2009, 4: 404-405.

[51] YIE Z, ZIELKE M A, BURGNER C B, et al. Comparison of parametric and linear mass detection in the presence of detection noise[J]. J. Micromech. Microeng., 2011, 21: 025027.

[52] CLELAND A N, ROUKES M L. Noise processes in nanomechanical resonators[J]. J. Appl. Phys., 2002, 92: 2758-2769.

[53] JIANG C, CUI Y S, ZHU K D. Ultrasensitive nanomechanical mass sensor using hybrid opto-electromechanical systems[J]. Opt. Express, 2014, 22: 13773-13783.

第6章 基于三腔光力系统的单向放大器

6.1 基于包含增益的三腔光力系统的单向放大器

6.1.1 引言

腔光力学领域研究电磁腔和机械振子通过辐射压形成的非线性相互作用[1-3]，其中有效光力耦合强度可以通过对腔施加强驱动而大大增强。当腔被驱动至红边带时，光力系统中取得了很多重要的进展，包括机械振子的量子基态冷却[4,5]、光力诱导透明[6-8]和量子态转化[9-11]等。此外，腔被驱动至蓝边带时，实验上观察到了量子纠缠[12,13]、微波放大[14-17]等现象。

另外，包括隔离器、循环器和单向放大器等在内的非互易性器件在通信和量子信息处理中起着重要的作用。为了实现非互易性，线性的非磁介质中电磁波方程满足的时间-反演对称性需要打破[18]。传统的打破时间-反演对称性的方法主要是利用磁光效应（如法拉第进动）[19-21]，该方法拥有体积大、价格高以及不利于集成等缺点。最近，几种新的方法被提出实现非互易性光学器件，包括折射率的动态时空调制[22,23]、光子和声子系统中的角动量偏置[24-26]、光学非线性[27-29]等。此外，研究人员利用一系列序参数泵浦在超导微波电路中实现了一种可集成的约瑟夫森循环器和单向放大器[30,31]。

最近，光力系统中腔和力学模式之间的辐射压诱导的参数耦合被用来打破时间-反演对称性[32-38]，引起了人们对光力系统中非互易性研究的热情。光力隔离器、循环器和单向放大器等非互易性器件理论上被相继提出[39,40]并在实验上得以实现[41-45]。这些工作大多数利用调控施加到腔模上的泵浦场的相对相位来实现非互易性。梅特尔曼（Metelamann）和科拉克（Clerk）提出了一种库工程方法来实现非互易性传输和放大[34]，该方法主要通过调节系统和耗散库之间的相互作用,最近被应用于光机械晶体电路中实现单向放大器[44]。此外,理论研究表明,双腔光力系统在包含机械增益[46]或者引入附加的机械驱动[47]时也可以实现单向放大。

这里我们提出一种基于包含光学增益的三腔光力系统实现不同频率的光场之间的单向放大的方案,其中一个光学腔掺杂了光学增益介质而另外两个耗散腔通过跳跃相互作用直接耦合起来,并且三个腔分别耦合于一个共同的机械振子。这个模型类似于文献[36]的模型,那里主要研究了没有增益介质时微波和光学光子之间的非互易性量子态转化。通过引入光学增益,研究人员在增益和损耗互相平衡的宇称-时间-对称的微腔系统中观察到了非互易性现象[28,48,49]。之后,包含光学增益的光力系统获得迅速的进展,包括声子激光[50,51]、光力诱导透明[52]、全光光子输运开关[53]、宇称-时间-对称性破缺的混沌[54]、增强的机械振子量子基态冷却[55]、探测机械运动时增强的灵敏度[56]等。本节的研究表明包含光学增益的光力系统可以用来实现微波场和光场之间的单向放大,并且放大的方向可以通过有效光力耦合强度之间的相位差进行控制。不同于之前的工作[40,41,45],本节的研究发现光学增益是实现放大的原因,而不是蓝失谐的泵浦场[40,41,45]或者附加的机械驱动[47]。此外,通过增加光力协同性可以极大抑制机械噪声的影响。值得指出的是,该光力系统中增益腔和损耗腔之间不存在直接的相互作用,因此可以用来实现不同频率之间的单向放大。最后,本节简单讨论了放大的信号场的群速度延迟,发现延迟时间相比于没有光学增益时可以明显提高。

6.1.2 模型

考虑的光力系统如图 6.1.1 所示。三个腔模 a_1、a_2、a_3 通过入射压力耦合于一个共同的力学模式 b。腔 a_1 的共振频率可以与腔 a_2 和 a_3 的频率相差很大,比如腔 a_1 是一个光学微腔而腔 a_2 和 a_3 是微波腔。这里我们考虑腔 a_1 是一个增益腔,可以通过在腔中掺杂 Er^{3+} 的硅并利用 1460 nm 的泵浦场对 Er^{3+} 进行泵浦从而发出 1550 nm 的光子[48]。此外,耗散腔 a_2 通过跳跃相互作用直接耦合于耗散腔 a_3,便于实现增益腔和耗散腔之间的单向放大。我们对每个腔都施加一个强的驱动场从而建立参数耦合。该光力系统的哈密顿量可表示为

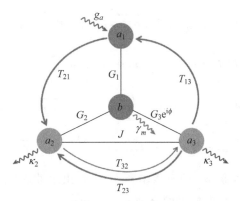

图 6.1.1　包含增益的光力系统示意图

（请扫Ⅱ页二维码看彩图）

$$H = \sum_{k=1}^{3} \omega_k a_k^+ a_k + \sum_{k=1}^{3} g_k a_k^+ a_k (b^+ + b) + \omega_m b^+ b + J(a_2^+ a_3 + a_3^+ a_2) + \\ \sum_{k=1}^{3} \left(\varepsilon_k a_k e^{i\omega_{d,k} t} + \varepsilon_k^* a_k^+ e^{-i\omega_{d,k} t} \right) \quad (6.1.1)$$

式中，$a_k(a_k^+)$ 是共振频率为 ω_k 的腔模 $a_k(k=1,2,3)$ 的湮灭（产生）算符，$b(b^+)$ 是共振频率为 ω_m 的力学模式 b 的湮灭（产生）算符，g_k 是腔模 a_k 和力学模式 b 之间的单光子光力耦合强度。第四项表示腔 a_2 和腔 a_3 之间的相互作用。最后一项表示强度为 ε_k、频率为 $\omega_{d,k}$ 的驱动场和腔模之间的相互作用。我们可以将腔模的每个算符写成经典平均值和量子涨落算符和的形式，即 $a_k \to \alpha_k e^{-i\omega_{d,k} t} + a_k$，其中经典的振幅 α_k 可以通过求解经典运动方程得到[43]。进一步变换到关于 $H_0 = \sum_{k=1}^{3} \omega_k a_k^+ a_k + \omega_m b^+ b$ 的旋转框架下，并在 $|\alpha_k| \gg 1$ 时忽略反旋和高阶项，可以得到如下线性化的哈密顿量：

$$H = G_1(a_1^+ b + a_1 b^+) + G_2(a_2^+ b + a_2 b^+) + G_3(a_3^+ b e^{-i\phi} + a_3 b^+ e^{i\phi}) + \\ J(a_2^+ a_3 + a_3^+ a_2) \quad (6.1.2)$$

式中，$G_k = g_k |\alpha_k| (k=1,2,3)$ 是力学模式和腔模 a_k 之间的有效耦合强度，α_k 的相位已经通过重新定义算符 a_k 和 b 的方式被吸收掉，只有它们之间的相位差 ϕ 有物理意义。这里假定了 $\omega_2 = \omega_3, \omega_{d,2} = \omega_{d,3}, \Delta_k = \omega_k - \omega_{d,k} = \omega_m$。

根据海森伯运动方程并增加相应的衰减和噪声项，可以得到如下量子朗之万方程[17,42,43]：

$$\dot{a}_1 = -\mathrm{i}G_1 b + \frac{g_a}{2}a_1 + \sqrt{\kappa_{\mathrm{ex},1}}\,a_{1,\mathrm{in}} + \sqrt{\kappa_{0,1}}\,a_{1,\mathrm{in}}^{(0)} + \sqrt{g}\,a_{1,\mathrm{in}}^{(g)} \quad (6.1.3)$$

$$\dot{a}_2 = -\mathrm{i}G_2 b - \mathrm{i}Ja_3 - \frac{\kappa_2}{2}a_2 + \sqrt{\kappa_{\mathrm{ex},2}}\,a_{2,\mathrm{in}} + \sqrt{\kappa_{0,2}}\,a_{2,\mathrm{in}}^{(0)} \quad (6.1.4)$$

$$\dot{a}_3 = -\mathrm{i}G_3 b\mathrm{e}^{-\mathrm{i}\phi} - \mathrm{i}Ja_2 - \frac{\kappa_3}{2}a_3 + \sqrt{\kappa_{\mathrm{ex},3}}\,a_{3,\mathrm{in}} + \sqrt{\kappa_{0,3}}\,a_{3,\mathrm{in}}^{(0)} \quad (6.1.5)$$

$$\dot{b} = -\mathrm{i}G_1 a_1 - \mathrm{i}G_2 a_2 - \mathrm{i}G_3 a_3 \mathrm{e}^{\mathrm{i}\phi} - \frac{\gamma_m}{2}b + \sqrt{\gamma_m}\,b_{\mathrm{in}} \quad (6.1.6)$$

其中,$g_a = g - \kappa_1$ 是腔 a_1 的有效增益率,g 是由泵浦腔 a_1 中 Er^{3+} 带来的增益[28,48]。腔 $a_k (k=1,2,3)$ 总的衰减率为 $\kappa_k = \kappa_{\mathrm{ex},k} + \kappa_{0,k}$,其中 $\kappa_{\mathrm{ex},k}$ 和 $\kappa_{0,k}$ 分别是外部衰减率和内在衰减率。γ_m 是力学模式 b 的衰减率。通过引入算符的傅里叶变换

$$o(\omega) = \int_{-\infty}^{+\infty} o(t)\mathrm{e}^{\mathrm{i}\omega t}\,\mathrm{d}t \quad (6.1.7)$$

$$o^+(\omega) = \int_{-\infty}^{+\infty} o^+(t)\mathrm{e}^{\mathrm{i}\omega t}\,\mathrm{d}t \quad (6.1.8)$$

与腔 a_1 增益相关的平均值为零的噪声算符 $a_{1,\mathrm{in}}^{(g)}$ 和 $a_{1,\mathrm{in}}^{(g)+}$ 满足关联函数[51,55,57,58]

$$\langle a_{1,\mathrm{in}}^{(g)}(\omega) a_{1,\mathrm{in}}^{(g)+}(\Omega) \rangle = 0,\ \langle a_{1,\mathrm{in}}^{(g)+}(\Omega) a_{1,\mathrm{in}}^{(g)}(\omega) \rangle = 2\pi\delta(\omega+\Omega) \quad (6.1.9)$$

这里我们假定了腔 a_1 的热光子占据数为零,原因是光频范围内 $\hbar\omega_1/k_\mathrm{B}T_e \gg 1$,其中 k_B 是玻耳兹曼常数,T_e 是环境的温度。

此外,通过外部耦合入射到腔 $a_k (k=1,2,3)$ 的输入场 $a_{k,\mathrm{in}}$ 满足关联函数[33,39,59]

$$\begin{cases} \langle a_{k,\mathrm{in}}(\omega) a_{k,\mathrm{in}}^+(\Omega) \rangle = 2\pi[s_{k,\mathrm{in}}(\omega)+1]\delta(\omega+\Omega) \\ \langle a_{k,\mathrm{in}}^+(\Omega) a_{k,\mathrm{in}}(\omega) \rangle = 2\pi s_{k,\mathrm{in}}(\omega)\delta(\omega+\Omega) \end{cases} \quad (6.1.10)$$

方程(6.1.10)中的 1 来源于真空噪声的影响,$s_{k,\mathrm{in}}(\omega)$ 表示的是通过外部耦合入射到腔 a_k 的弱的探测场。噪声算符 $a_{k,\mathrm{in}}^{(0)}$ 和 b_{in} 分别对应于腔 a_k 的内部损耗和机械振子的衰减。在白噪声近似下,以上平均值为零的噪声算符满足如下非零关联函数[17,42,43]:

$$\begin{cases} \langle a_{k,\mathrm{in}}^{(0)}(\omega) a_{k,\mathrm{in}}^{(0)+}(\Omega) \rangle = 2\pi\delta(\omega+\Omega) \\ \langle b_{\mathrm{in}}(\omega) b_{\mathrm{in}}^+(\Omega) \rangle = 2\pi(n_m+1)\delta(\omega+\Omega) \\ \langle b_{\mathrm{in}}^+(\Omega) b_{\mathrm{in}}(\omega) \rangle = 2\pi n_m \delta(\omega+\Omega) \end{cases} \quad (6.1.11)$$

这里假定了腔的库温度极低时腔场的热光子占据数为零[1,40,42]，机械振子的热声子数 $n_m = 1/[\exp(\hbar\omega_m/k_B T_e) - 1]$。

方便起见，量子朗之万方程(6.1.4)～方程(6.1.6)可以写成矩阵形式

$$\dot{\boldsymbol{\mu}} = \boldsymbol{M}\boldsymbol{\mu} + \boldsymbol{L}\boldsymbol{\mu}_{\text{in}} \tag{6.1.12}$$

其中矢量 $\boldsymbol{\mu} = (a_1, a_2, a_3, b)^T$，$\boldsymbol{\mu}_{\text{in}} = (a_{1,\text{in}}, a_{2,\text{in}}, a_{3,\text{in}}, a_{1,\text{in}}^{(0)}, a_{2,\text{in}}^{(0)}, a_{3,\text{in}}^{(0)}, a_{1,\text{in}}^{(g)}, b_{\text{in}})^T$，T 代表转置，系数矩阵

$$\boldsymbol{M} = \begin{pmatrix} \frac{g_a}{2} & 0 & 0 & -iG_1 \\ 0 & -\kappa_2/2 & -iJ & -iG_2 \\ 0 & -iJ & -\kappa_3/2 & -iG_3 e^{-i\phi} \\ -iG_1 & -iG_2 & -iG_3 e^{i\phi} & -\gamma_m/2 \end{pmatrix} \tag{6.1.13}$$

$$\boldsymbol{L} = \begin{pmatrix} \sqrt{\kappa_{\text{ex},1}} & 0 & 0 & \sqrt{\kappa_{0,1}} & 0 & 0 & \sqrt{g} & 0 \\ 0 & \sqrt{\kappa_{\text{ex},2}} & 0 & 0 & \sqrt{\kappa_{0,2}} & 0 & 0 & 0 \\ 0 & 0 & \sqrt{\kappa_{\text{ex},3}} & 0 & 0 & \sqrt{\kappa_{0,3}} & 0 & 0 \\ 0 & 0 & 0 & 0 & 0 & 0 & 0 & \sqrt{\gamma_m} \end{pmatrix}$$

$$\tag{6.1.14}$$

该系统只有当矩阵 $\boldsymbol{M}$ 的所有本征值的实部都为负数时才稳定。稳定性条件可以根据劳斯-赫尔维茨(Routh-Hurwitz)判据[60,61]推导出，由于具体形式过于复杂，这里不再给出。但是，接下来我们会在数值上检查稳定性条件并在稳定区域中选择参数。

频域空间中方程(6.1.12)的解为

$$\boldsymbol{\mu}(\omega) = -(\boldsymbol{M} + i\omega\boldsymbol{I})^{-1}\boldsymbol{L}\boldsymbol{\mu}_{\text{in}}(\omega) \tag{6.1.15}$$

其中 $\boldsymbol{I}$ 代表单位矩阵。将方程(6.1.15)代入到标准的输入-输出关系 $\boldsymbol{\mu}_{\text{out}}(\omega) = \boldsymbol{\mu}_{\text{in}}(\omega) - \boldsymbol{L}^T\boldsymbol{\mu}(\omega)$，可以得到

$$\boldsymbol{\mu}_{\text{out}}(\omega) = \boldsymbol{T}(\omega)\boldsymbol{\mu}_{\text{in}}(\omega) \tag{6.1.16}$$

其中输出场矢量 $\boldsymbol{\mu}_{\text{out}}(\omega)$ 是 $\boldsymbol{\mu}_{\text{out}} = (a_{1,\text{out}}, a_{2,\text{out}}, a_{3,\text{out}}, a_{1,\text{out}}^{(0)}, a_{2,\text{out}}^{(0)}, a_{3,\text{out}}^{(0)}, a_{1,\text{out}}^{(g)}, b_{\text{out}})^T$ 的傅里叶变换，透射矩阵

$$\boldsymbol{T}(\omega) = \boldsymbol{I} + \boldsymbol{L}^T(\boldsymbol{M} + i\omega\boldsymbol{I})^{-1}\boldsymbol{L} \tag{6.1.17}$$

矩阵元 $\boldsymbol{T}_{ij}(\omega)(i,j = 1,2,3)$ 代表入射到腔 a_j 上的信号从腔 a_i 输出时的透射振幅。

6.1.3 单向放大器

本节讨论与腔共振的输入探测场如何在腔模 a_1 和 a_2 之间实现单向放大。根据方程(6.1.13)和方程(6.1.17),可以得到透射矩阵元

$$T_{12}(\omega) = -\frac{\sqrt{\eta_1 \eta_2 \kappa_1 \kappa_2}}{A(\omega)} G_1 (G_2 \Gamma_3 + \mathrm{i} J G_3 \mathrm{e}^{\mathrm{i}\phi}) \qquad (6.1.18)$$

$$T_{21}(\omega) = -\frac{\sqrt{\eta_1 \eta_2 \kappa_1 \kappa_2}}{A(\omega)} G_1 (G_2 \Gamma_3 + \mathrm{i} J G_3 \mathrm{e}^{-\mathrm{i}\phi}) \qquad (6.1.19)$$

式中,

$$A(\omega) = \Gamma_1 (\Gamma_2 \Gamma_3 \Gamma_m + \Gamma_3 G_2^2 + \Gamma_2 G_3^2 + \Gamma_m J^2 + 2\mathrm{i} G_2 G_3 J \cos\phi) + G_1^2 (\Gamma_2 \Gamma_3 + J^2)$$

$\Gamma_1 = g_a/2 + \mathrm{i}\omega, \Gamma_2 = -\kappa_2/2 + \mathrm{i}\omega, \Gamma_3 = -\kappa_3/2 + \mathrm{i}\omega, \Gamma_m = -\gamma_m/2 + \mathrm{i}\omega$

$\eta_k = \kappa_{\mathrm{ex},k}/\kappa_k (k=1,2,3)$

是腔 a_k 的耦合效率[37,38,43]。

为了能够实现单向放大器,要求从腔 a_1 输入的探测场从腔 a_2 输出时可以被放大,但是从腔 a_2 输入的探测场不能从腔 a_1 输出,即 $|T_{21}|^2 > 1$, $|T_{12}|^2 = 0$。从方程(6.1.18)和方程(6.1.19)可以看出,如果 $\phi = -\pi/2, G_3 = G_2 \kappa_2 / 2J$, 那么 $|T_{21}(0)| \neq 0, |T_{12}(0)| = 0$。接下来会证实腔 a_1 的光学增益会导致 $|T_{21}|^2$ 大于 1。此外,为了能够阻止输入场传输到 a_3 和 b 等其他模式引起损失,要求 $|T_{12}|^2 = 0$ 时 $|T_{i1}/T_{21}| \ll 1 (i \neq 2)$。通过选择 $J = \sqrt{\kappa_2 \kappa_3}/2$, 可以得到 $|T_{31}| = 0$ 和 $|T_{41}/T_{21}| = \sqrt{\gamma_m \kappa_{\mathrm{ex},2}}/2G_2 \ll 1$。因此,与腔共振的入射探测场从腔 a_1 到腔 a_2 实现单向放大的条件包括

$$\phi = -\pi/2, \quad G_3 = G_2 \kappa_2 / 2J, \quad J = \sqrt{\kappa_2 \kappa_3}/2 \qquad (6.1.20)$$

在上述条件下,共振时的透射振幅 T_{21} 可以简化为

$$T_{21}(0) = \frac{8\sqrt{\eta_1 \eta_2 \kappa_1 \kappa_2} G_1 G_2}{4\kappa_2 G_1^2 - 4g_a G_2^2 - g_a \kappa_2 \gamma_m} = \frac{2\sqrt{\eta_1 \eta_2 C_1 C_2} \kappa_1/g_a}{C_1 \kappa_1/g_a - C_2 - 1} \qquad (6.1.21)$$

其中光力协同性 $C_k = 4G_k^2/\kappa_k \gamma_m (k=1,2)$。腔 a_1 的有效增益率 g_a 可以通过调节增益率 g 进行控制,简单起见可以假设 $g_a = \kappa_1$。放大器的增益率[40]

$$G = |T_{21}(0)|^2 = \frac{4\eta_1 \eta_2 C_1 C_2}{(C_1 - C_2 - 1)^2} \qquad (6.1.22)$$

为了能够基于包含光学增益的光力系统实现单向放大器,该系统应该工作在稳定区域。图 6.1.2 给出了稳定图随耦合强度 G_1 和 G_2 变化的情况,用到的参数为 $g_a/(2\pi)=\kappa_1/(2\pi)=\kappa_2/(2\pi)=2\,\mathrm{MHz},\kappa_3/(2\pi)=2\,\mathrm{MHz},\gamma_m=\kappa_2/100,\phi=-\pi/2,G_3=G_2\kappa_2/(2J),J=\sqrt{\kappa_2\kappa_3}/2$。从图中可以看出,系统由腔 a_1 的光学增益只在一个较窄的区域保持稳定。此外,当耦合强度 G_1 大于某一临界值($G_1/(2\pi)\geqslant 1.1\,\mathrm{MHz}$),只要耦合强度 G_2 比 G_1 小一点系统即可处于稳定区域。因此,对于给定的 G_1,可以选取 $\widetilde{G}_2=\widetilde{G}_1-0.1\sqrt{\widetilde{G}_1}$ 以确保系统是稳定的,其中 $\widetilde{G}_1(\widetilde{G}_2)$ 是 $G_1(G_2)$ 以 MHz 为单位时对应的无量纲数值。

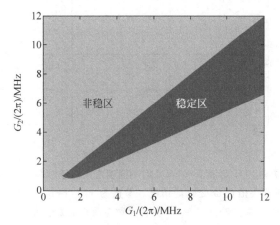

图 6.1.2　随有效光力耦合强度 G_1 和 G_2 变化的稳定图

(请扫 Ⅱ 页二维码看彩图)

根据透射矩阵,可以研究透射率对频率和相位差的依赖关系。图 6.1.3 给出了相位差 $\phi=\pi/2$ 和 $\phi=-\pi/2$ 时不同腔之间的透射率 $|T_{ij}|^2(i,j=1,2,3)$ 随探测失谐量 ω 变化的情况。当相位差被调至 $\phi=-\pi/2$,从图中可以看出 $\omega=0$ 处 $|T_{21}|^2$ 达到其最大值 30 dB,而 $|T_{12}|^2\approx 0$。因此,入射到腔 a_1 上的信号从腔 a_2 出射时可以被放大很多,但是入射到腔 a_2 上的信号不能从腔 a_1 出射。导致这种单向放大的原因是两条不同路径之间的干涉效应,其中一条路径是 $a_1\rightarrow b\rightarrow a_2$,另一条路径是 $a_1\rightarrow b\rightarrow a_3\rightarrow a_2$。当相位差 $\phi=-\pi/2$ 时,两条路径之间的相长干涉以及腔 a_1 的光学增益导致了信号从腔 a_1 到腔 a_2 的单向放大,而相反方向的传输由于相消干涉被抑制($|T_{12}|=0$)。此外,入射到

腔 a_3 上的信号可以被单向放大到腔 a_1($|T_{13}|^2 \approx 30$ dB,$|T_{31}|^2 \approx 0$)。因此,基于该光力系统可以实现一种单向放大器,输入的信号可以沿着路径 $a_3 \to a_1 \to a_2$ 被单向放大。另一方面,如果相位 ϕ 从 $-\pi/2$ 调节成 $\pi/2$ 时,信号将会沿着相反的方向(即 $a_2 \to a_1 \to a_3$)被放大。

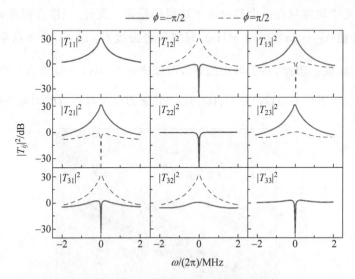

图 6.1.3 相位差 $\phi = \pi/2$ 和 $\phi = -\pi/2$ 时不同腔之间的透射率 $|T_{ij}|^2$($i,j = 1,2,3$) 随探测失谐量 ω 变化的情况。参数为 $\eta_{1,2,3} = 1, \kappa_1/(2\pi) = g_a/(2\pi) = 2$ MHz,$G_1/(2\pi) = 2$ MHz,其他参数与图 6.1.2 相同

(请扫 Ⅱ 页二维码看彩图)

我们主要研究腔 a_1 和腔 a_2 之间的单向放大。图 6.1.4 给出了 $\omega = 0$ 时透射率 $|T_{12}|^2$ 和 $|T_{21}|^2$ 随相位差 ϕ 变化的情况。从图中可以看出,当 $\phi = 0$,$\pm\pi$ 时 $|T_{12}|^2 = |T_{21}|^2$,因此满足洛伦兹互易性定理,并且光力系统对信号场的响应是互易的。但是,当 $\phi \neq n\pi$(n 是一个整数)时,时间-反演-对称性破缺,光力系统呈现出非互易性响应。当 $-\pi < \phi < 0$ 时,我们发现 $|T_{12}|^2 < |T_{21}|^2$,但是当 $0 < \phi < \pi$ 时,$|T_{12}|^2 > |T_{21}|^2$。在给定参数下,最大非互易性响应出现在 $\phi = -\pi/2$($|T_{21}|^2 \approx 30$ dB,$|T_{12}|^2 \approx 0$)和 $\phi = \pi/2$($|T_{12}|^2 \approx 30$ dB,$|T_{21}|^2 \approx 0$)。因此,通过调节相位差 ϕ 可以基于该系统实现单向放大器。

接下来研究腔 a_1 的有效增益率 g_a 对透射率 $|T_{21}|^2$ 的影响。图 6.1.5 给出了不同增益率 g_a 时透射率 $|T_{21}|^2$ 随探测失谐量 ω 变化的情况。当

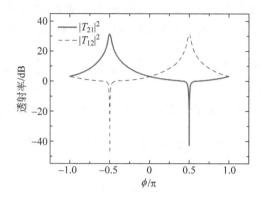

图 6.1.4 $\omega=0$ 时透射率 $|T_{12}|^2$ 和 $|T_{21}|^2$ 随相位差 ϕ 变化的曲线,参数与图 6.1.3 相同

(请扫Ⅱ页二维码看彩图)

$g_a/(2\pi)$ 从 2 MHz 减小到 0.5 MHz 时,共振处最大透射率 $|T_{21}|^2$ 从 35 dB 减小到 8 dB,而 $|T_{12}|^2$ 始终为零。此外,如果增益腔变成耗散腔,比如 $g_a/(2\pi)=-2$ MHz,共振处的透射率 $|T_{21}|^2 \approx 1$,表示出现了光力诱导透明现象[6,7]。在这种情况下,非互易性传输现象在该光力系统中仍然存在[36],但是探测场不能被放大。因此,光学增益是放大的物理原因,并且相位差可以用来控制该系统中非互易性传输现象。

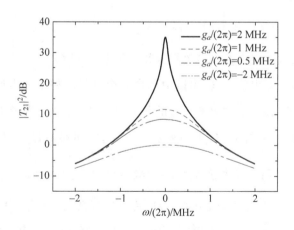

图 6.1.5 不同增益率 g_a 时透射率 $|T_{21}|^2$ 随探测失谐量 ω 变化的情况。参数除 $\phi=-\pi/2$, $G_1/(2\pi)=5$ MHz 外与图 6.1.3 相同

(请扫Ⅱ页二维码看彩图)

对于相位保持的线性放大器,带宽通常随着增益的变大而减小,这从图 6.1.5 中也可以看出。放大器的带宽可以从方程(6.1.19)中 $T_{21}(\omega)$ 的分母近似得到。如果假设 $g_a = \kappa_2 = \kappa_3 = \kappa$,那么 $A(\omega) \approx \frac{1}{8}\kappa^3 \gamma_m (C_1 - C_2 - 1) + \frac{1}{4}\gamma_m \kappa^2 (\kappa/\gamma_m - C_1) i\omega$,其中只保留了 ω 的一阶项[40]。带宽 Γ 可以由满足 $2|A(0)|^2 = |A(\omega)|^2$ 的最小的 $|\omega|$ 值得到,近似表达式为

$$\Gamma = \left| \frac{\kappa(C_1 - C_2 - 1)}{\kappa/\gamma_m - C_1} \right| \tag{6.1.23}$$

在这个包含光学增益的光力系统,可以从图 6.1.2 看出光力协同性较小时系统是不稳定的。因此,我们考虑光力协同性较大的情况,即 $C_1 > C_2 \gg \kappa/\gamma_m$,则带宽 $\Gamma = \kappa(C_1 - C_2 - 1)/C_1$,增益带宽积 $P \equiv \Gamma\sqrt{G} \to 2\kappa$。

放大器增加的噪声数可以通过计算腔 a_2 的输出谱得到,表示如下[40,62]:

$$S_{2,\text{out}}(\omega) = \frac{1}{2}\int \frac{d\Omega}{2\pi} \langle a_{2,\text{out}}(\omega) a_{2,\text{out}}^+(\Omega) + a_{2,\text{out}}^+(\Omega) a_{2,\text{out}}(\omega) \rangle$$

$$= \sum_{i=1}^{3}\left[s_{i,\text{in}}(\omega) + \frac{1}{2} \right] |T_{2i}(\omega)|^2 + \frac{1}{2}\sum_{i=4}^{7} |T_{2i}(\omega)|^2 +$$

$$\left(n_m + \frac{1}{2} \right) |T_{28}(\omega)|^2 \tag{6.1.24}$$

其中利用了频域空间中的噪声关联函数和关系式 $o^+(\omega) = [o(-\omega)]^+$。从方程(6.1.24)可以看出腔 a_2 输出谱中包含 8 个分量。如果考虑通过外部耦合入射到腔 a_1 而从腔 a_2 输出的信号的单向放大,即 $|T_{21}|^2$,那么与热声子占据数 n_i 相关的其他透射率 $|T_{2i}|^2 (i = 2, 3, 4, \cdots, 8)$ 可被看成是噪声,并且假设 $s_{2,\text{in}} = s_{3,\text{in}} = 0$。因此,增加到放大器的噪声可定义为 $N_2(\omega) = G^{-1}\sum_{i=2}^{8}(n_i + 1/2)|T_{2i}|^2$ [40,63-65]。$N_2(\omega)$ 的一般形式过于复杂,这里就不再给出。但是,如果 $\eta_{1,2,3} = 1, g_a = \kappa_1, \omega = 0$,腔 a_2 输出场中增加的噪声可由下式给出:

$$N_2(0) = \frac{1}{2}\frac{(C_1 + C_2 - 1)^2}{4C_1 C_2} + \left(n_m + \frac{1}{2} \right)\frac{1}{C_1} + 1 \tag{6.1.25}$$

其中已经假定了腔场的热光子占据数 $n_i = 0 (i = 1, 2, 3, \cdots, 7)$,机械振子的热

声子占据数 $n_s = n_m$。方程(6.1.25)右边最后一项 1 主要来源于腔 a_1 的增益。可以从方程(6.1.25)看出机械振子的热噪声可以通过增加光力协同性 C_1 进行抑制。当 C_1 和 C_2 较大时,得到零频时 $N_2(0) \to 1.5$。

图 6.1.6(a)和(b)分别给出了不同光力耦合强度时透射率$|T_{21}|^2$和增加的噪声随探测失谐量 ω 变化的情况。从图 6.1.6(a)可以看出$|T_{21}|^2$的峰值随着 $G_1/(2\pi)$ 从 2 MHz 增加到 10 MHz 而不断增大。此外,图 6.1.6(b)表明增加的噪声数 $N_2(\omega)$ 随着光力耦合常数 G_1 和 G_2 增加而不断减小。尤其是当 $G_{1,2}$ 足够大时,$G_{1,2} \gg 1$ 并且共振时的 $N_2(0)$ 可以接近于 1.5。

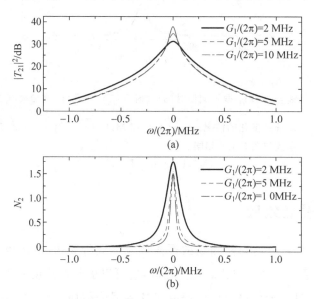

图 6.1.6 不同光力耦合强度时(a)透射率$|T_{21}|^2$和(b)增加的噪声随探测失谐量 ω 变化的曲线。$G_1/(2\pi) = 2$ MHz、5 MHz、10 MHz,$\widetilde{G}_2 = \widetilde{G}_1 - 0.1\sqrt{\widetilde{G}_1}$。其他参数除 $\phi = -\pi/2, n_m = 100$ 外与图 6.1.3 相同

(请扫 II 页二维码看彩图)

此外,图 6.1.7 讨论了腔的内在损耗率对放大器的增益和增加噪声的影响。我们发现当耦合率 η_k 从 1 减小为 0.5(临界耦合)时,放大器的增益不断减小,这也可以从方程(6.1.21)看出。同时,图 6.1.7(b)表明随着耦合率的减小,放大器增加的噪声不断增大。因此,每个腔都有高的耦合率 η_k,对增加放大器的增益和抑制增加的噪声都是有帮助的。

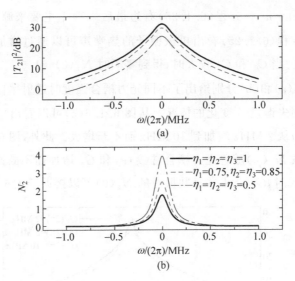

图 6.1.7 耦合效率 η_k 取不同值时(a)透射率 $|T_{21}|^2$ 及(b)增加噪声 N_2 随失谐量 ω 变化的情况。$G_1/(2\pi)=2\ \mathrm{MHz}, \widetilde{G}_2=\widetilde{G}_1-0.1\sqrt{\widetilde{G}_1}$,其他参数与图 6.1.6 相同

(请扫Ⅱ页二维码看彩图)

6.1.4 慢光效应

最后,我们研究透射探测场的慢光效应。众所周知,电磁诱导透明窗口中的探测场通常伴随着快速的相位色散,这可以引起群速度的急速下降。在光力系统中,过去十年已经广泛研究了这种慢光效应[7,8]。本节我们研究单向放大伴随的慢光效应。透射探测场中的光学群速度延迟定义为[7]

$$\tau = \frac{\mathrm{d}\theta}{\mathrm{d}\omega} \tag{6.1.26}$$

其中,$\theta=\arg[T_{21}(\omega)]$ 是腔 a_2 的输出场中探测场频率分量对应的相位。

图 6.1.8 给出了从腔 a_2 透射的探测场的相位(图 6.1.8(a))和群速度延迟(图 6.1.8(b))变化的情况。从图 6.1.8(a)可以看出,放大的透射探测场伴随着快速的相位色散,并且在 $\omega=0$ 处的斜率随着耦合强度 G_1 增加而变大。因此,随着 G_1 增加最大群速度延迟也变大,可从图 6.1.8(b)看出。此外,图 6.1.8(b)的插图所示是腔 a_1 也是损耗腔时群速度延迟 τ 随着探测失谐量

ω 变化的情况。在这种情况下,透射的探测场中可以出现光力诱导透明现象,如图 6.1.5 所示。当 $G_1/(2\pi)=5$ MHz 时,可以看出对于增益腔($g_a/(2\pi)=2$ MHz)而言共振时最大群速度延迟大约是 4.3 μs,而损耗腔时($g_a/(2\pi)=-2$ MHz)最大群速度延迟大约是 0.16 μs。因此,包含光学增益的光力系统可以出现单向放大,并且群速度延迟可以被进一步延长。

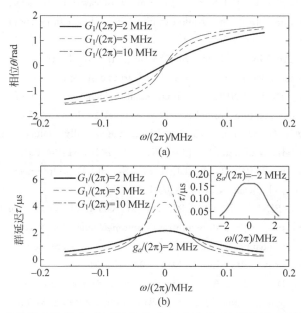

图 6.1.8　有效耦合强度 G_1 取不同数值时(a)相位和(b)群速度延迟随探测失谐量 ω 变化的情况。参数为 $g_a/(2\pi)=2$ MHz,图(b)插图中 $g_a/(2\pi)=-2$ MHz,$G_1/(2\pi)=5$ MHz,其他参数与图 6.1.5 相同

(请扫Ⅱ页二维码看彩图)

6.1.5　小结

本节研究了如何基于包含增益的光力系统实现单向放大器[66]。增益腔和损耗腔之间的透射可以单向放大并且放大的方向依赖于有效光力耦合强度之间的相位差。最大的放大器增益可以通过增加增益腔的增益率和有效光力耦合强度进行提高。在光力协同性较大时,腔输出场中机械噪声的影响可以被极大抑制。此外,在包含增益的光力系统中透射探测场的群速度延迟相比于没有增益的光力系统中的群速度延迟可以提高一两个数量级。

参 考 文 献

[1] ASPELMEYER M, KIPPENBERG T J, MARQUARDT F. Cavity optomechanics [J]. Rev. Mod. Phys., 2014, 86: 1391.
[2] MARQUARDT F, GIRVIN S M. Optomechanics[J]. Physics, 2009, 2: 40.
[3] XIONG H, SI L G, LÜ X Y, et al. Review of cavity optomechanics in the weak-coupling regime: from linearization to intrinsic nonlinear interactions[J]. Sci. China Phys. Mech. Astron., 2015, 58: 1.
[4] CHAN J, ALEGRE T P M, SAFAVI-NAEINI A H, et al. Laser cooling of a nanomechanical oscillator into its quantum ground state[J]. Nature, 2011, 478: 89.
[5] TEUFEL J D, DONNER T, LI D, et al. Sideband cooling of micromechanical motion to the quantum ground state[J]. Nature, 2011, 475: 359.
[6] AGARWAL G S, HUANG S. Electromagnetically induced transparency in mechanical effects of light[J]. Phys. Rev. A, 2010, 81: 041803.
[7] WEIS S, RIVIÈRE R, DELÉGLISE S, et al. Optomechanically induced transparency[J]. Science, 2010, 330: 1520.
[8] SAFAVI-NAEINI A H, ALEGRE T P M, CHAN J, et al. Electromagnetically induced transparency and slow light with optomechanics[J]. Nature, 2011, 472: 69.
[9] TIAN L. Adiabatic state conversion and pulse transmission in optomechanical systems[J]. Phys. Rev. Lett., 2012, 108: 153604.
[10] WANG Y D, CLERK A A. Using interference for high fidelity quantum state transfer in optomechanics[J]. Phys. Rev. Lett., 2012, 108: 153603.
[11] ANDREWS R W, PETERSON R W, PURDY T P, et al. Bidirectional and efficient conversion between microwave and optical light[J]. Nat. Phys., 2014, 10: 321.
[12] TIAN L. Robust photon entanglement via quantum interference in optomechanical interfaces[J]. Phys. Rev. Lett., 2013, 110: 233602.
[13] WANG Y D, CLERK A A. Reservoir-engineered entanglement in optomechanical systems[J]. Phys. Rev. Lett., 2013, 110: 253601.
[14] MASSEL F, HEIKKILÄ T T, PIRKKALAINEN J M, et al. Microwave amplification with nanomechanical resonators[J]. Nature, 2011, 480: 351.
[15] OCKELOEN-KORPPI C F, DAMSKÄGG E, PIRKKALAINEN J M, et al. Low-noise amplification and frequency conversion with a multiport microwave optomechanical device[J]. Phys. Rev. X, 2016, 6: 041024.
[16] OCKELOEN-KORPPI C F, DAMSKÄGG E, PIRKKALAINEN J M, et al. Noiseless quantum measurement and squeezing of microwave fields utilizing mechanical vibrations[J]. Phys. Rev. Lett., 2017, 118: 103601.

[17] TÓTH L D, BERNIER N R, NUNNENKAMP A, et al. A dissipative quantum reservoir for microwave light using a mechanical oscillator[J]. Nat. Phys., 2017, 13: 787.

[18] POTTON R J. Reciprocity in optics[J]. Rep. Prog. Phys., 2004, 67: 717.

[19] HALDANE F D M, RAGHU S. Possible realization of directional optical waveguides in photonic crystals with broken time-reversal symmetry[J]. Phys. Rev. Lett., 2008, 100: 013904.

[20] KHANIKAEV A B, MOUSAVI S H, SHVETS G, et al. One-way extraordinary optical transmission and nonreciprocal spoof plasmons[J]. Phys. Rev. Lett., 2010, 105: 126804.

[21] BI L, HU J, JIANG P, et al. On-chip optical isolation in monolithically integrated non-reciprocal optical resonators[J]. Nat. Photon., 2011, 5: 758.

[22] LIRA H, YU Z, FAN S, et al. Electrically driven nonreciprocity induced by interband photonic transition on a silicon chip[J]. Phys. Rev. Lett., 2012, 109: 033901.

[23] FANG K, YU Z, FAN S. Photonic Aharonov-Bohm effect based on dynamic modulation[J]. Phys. Rev. Lett., 2012, 108: 153901.

[24] FLEURY R, SOUNAS D L, SIECK C F, et al. Sound isolation and giant linear nonreciprocity in a compact acoustic circulator[J]. Science, 2014, 343: 516.

[25] ESTEP N A, SOUNAS D L, SORIC J, et al. Magnetic-free non-reciprocity and isolation based on parametrically modulated coupled-resonator loops[J]. Nat. Phys., 2014, 10: 923.

[26] WANG D W, ZHOU H T, GUO M J, et al. Optical diode made from a moving photonic crystal[J]. Phys. Rev. Lett., 2013, 110: 093901.

[27] FAN L, WANG J, VARGHESE L T, et al. An all-silicon passive optical diode [J]. Science, 2012, 335: 447.

[28] CHANG L, JIANG X, HUA S, et al. Parity-time symmetry and variable optical isolation in active-passive-coupled microresonators[J]. Nat. Photon., 2014, 8: 524.

[29] GUO X, ZOU C L, JUNG H, et al. On-chip strong coupling and efficient frequency conversion between telecom and visible optical modes[J]. Phys. Rev. Lett., 2016, 117: 123902.

[30] SLIWA K M, HATRIDGE M, NARLA A, et al. Reconfigurable Josephson circulator/directional amplifier[J]. Phys. Rev. X, 2015, 5: 041020.

[31] LECOCQ F, RANZANI L, PETERSON G A, et al. Nonreciprocal microwave signal processing with a field-programmable Josephson amplifier[J]. Phys. Rev. Appl., 2017, 7: 024028.

[32] HAFEZI M, RABL P. Optomechanically induced non-reciprocity in microring resonators[J]. Opt. Express, 2012, 20: 7672.

[33] XU X W, LI Y. Optical nonreciprocity and optomechanical circulator in three-mode optomechanical systems[J]. Phys. Rev. A, 2015, 91: 053854.

[34] METELMANN A, CLERK A A. Nonreciprocal photon transmission and amplification via reservoir engineering[J]. Phys. Rev. X, 2015, 5: 021025.

[35] SHEN Z, ZHANG Y L, CHEN Y, et al. Experimental realization of optomechanically induced non-reciprocity[J]. Nat. Photon., 2016, 10: 657.

[36] TIAN L, LI Z. Nonreciprocal quantum-state conversion between microwave and optical photons[J]. Phys. Rev. A, 2017, 96: 013808.

[37] MIRI M A, RUESINK F, VERHAGEN E, et al. Optical nonreciprocity based on optomechanical coupling[J]. Phys. Rev. Appl., 2017, 7: 064014.

[38] PETERSON G A, LECOCQ F, CICAK K, et al. Demonstration of efficient nonreciprocity in a microwave optomechanical circuit[J]. Phys. Rev. X, 2017, 7: 031001.

[39] XU X W, LI Y, CHEN A X, et al. Nonreciprocal conversion between microwave and optical photons in electro-optomechanical systems[J]. Phys. Rev. A, 2016, 93: 023827.

[40] MALZ D, TÓTH L D, BERNIER N R, et al. Quantum-limited directional amplifiers with optomechanics[J]. Phys. Rev. Lett., 2018, 120: 023601.

[41] RUESINK F, MIRI M A, ALÙ A, et al. Nonreciprocity and magnetic-free isolation based on optomechanical interactions[J]. Nat. Commun, 2016, 7: 13662.

[42] BERNIER N R, TÒTH L D, KOOTTANDAVIDA A, et al. Nonreciprocal reconfigurable microwave optomechanical circuit[J]. Nat. Commun., 2017, 8: 604.

[43] BARZANJEH S, WULF M, PERUZZO M, et al. Mechanical on-chip microwave circulator[J]. Nat. Commun., 2017, 8: 953.

[44] FANG K, LUO J, METELMANN A, et al. Generalized non-reciprocity in an optomechanical circuit via synthetic magnetism and reservoir engineering[J]. Nat. Phys., 2017, 13: 465.

[45] SHEN Z, ZHANG Y L, CHEN Y, et al. Reconfigurable optomechanical circulator and directional amplifier[J]. Nat. Commun., 2018, 9: 1797.

[46] ZHANG X Z, TIAN L, LI Y. Optomechanical transistor with mechanical gain[J]. Phys. Rev. A, 2018, 97: 043818.

[47] LI Y, HUANG Y Y, ZHANG X Z, et al. Optical directional amplification in a three-mode optomechanical system[J]. Opt. Express, 2017, 25: 18907.

[48] PENG B, ÖZDEMIR Ş K, LEI F, et al. Parity-time-symmetric whispering-gallery microcavities[J]. Nat. Phys., 2014, 10: 394.

[49] LUO X B, HUANG J H, ZHONG H H, et al. Pseudo-parity-time symmetry in optical systems[J]. Phys. Rev. Lett., 2013, 110: 243902.

[50] JING H, ÖZDEMIR S K, LÜ X Y, et al. PT-symmetric phonon laser[J]. Phys. Rev. Lett., 2014, 113: 053604.

[51] HE B, YANG L, XIAO M. Dynamical phonon laser in coupled active-passive microresonators[J]. Phys. Rev. A, 2016, 94: 031802.

[52] JING H, ÖZDEMIR S K, GENG Z, et al. Optomechanically-induced transparency in parity-time-symmetric microresonators[J]. Sci. Rep., 2015, 5: 9663.

[53] DU L, LIU Y M, ZHANG Y. All-optical photon switching, router and amplifier using a passive-active optomechanical system[J]. Europhys. Lett, 2018, 122: 24001.

[54] LÜ X Y, JING H, MA J Y, et al. PT-symmetry-breaking chaos in optomechanics [J]. Phys. Rev. Lett., 2015, 114: 253601.

[55] LIU Y L, LIU Y X. Energy-localization-enhanced ground-state cooling of a mechanical resonator from room temperature in optomechanics using a gain cavity [J]. Phys. Rev. A, 2017, 96: 023812.

[56] LIU Z P, ZHANG J, ÖZDEMIR S K, et al. Metrology with PT-symmetric cavities: enhanced sensitivity near the PT-phase transition[J]. Phys. Rev. Lett., 2016, 117: 110802.

[57] AGARWAL G S, QU K. Spontaneous generation of photons in transmission of quantum fields in PT-symmetric optical systems [J]. Phys. Rev. A, 2012, 85: 031802.

[58] KEPESIDIS K V, MILBURN T J, HUBER J, et al. PT-symmetry breaking in the steady state of microscopic gain-loss systems [J]. New J. Phys., 2016, 18: 095003.

[59] AGARWAL G S, HUANG S. Optomechanical systems as single-photon routers [J]. Phys. Rev. A, 2012, 85: 021801.

[60] DEJESUS E X, KAUFMAN C. Routh-Hurwitz criterion in the examination of eigenvalues of a system of nonlinear ordinary differential equations[J]. Phys. Rev. A, 1987, 35: 5288.

[61] GRADSHTEYN I S, RYZHIK I M. Table of integrals[M]. Series and Products: Academic Press, 1980.

[62] METELMANN A, CLERK A A. Quantum-limited amplification via reservoir engineering[J]. Phys. Rev. Lett., 2014, 112: 133904.

[63] CAVES C M. Quantum limits on noise in linear amplifiers[J]. Phys. Rev. D, 1982, 26: 1817.

[64] CLERK A A, DEVORET M H, GIRVIN S M, et al. Introduction to quantum noise, measurement, and amplification[J]. Rev. Mod. Phys., 2010, 82: 1155.

[65] NUNNENKAMP A, SUDHIR V, FEOFANOV A K, et al. Quantum-limited amplification and parametric instability in the reversed dissipation regime of cavity optomechanics[J]. Phys. Rev. Lett., 2014, 113: 023604.

[66] JIANG C, SONG L N, LI Y. Directional amplifier in an optomechanical system with optical gain[J]. Phys. Rev. A, 2018, 97: 053812.

6.2 基于三腔光力系统的量子极限的单向放大器

6.2.1 引言

介质中的电磁波传输在源和探测器相互交换时通常是不变的[1],但是在经典和量子信息处理中实现单向传输是一个基本要求。因此,包括隔离器、循环器和单向放大器在内的非互易性器件变得尤为重要,它们可以有效地保护信号源。实现非互易性传输的常用办法是通过施加外加磁场使信号的传输通道发生偏置,这就需要磁光材料,因此很难实现集成和小型化[2-4]。在过去几十年中,多种不需要磁光效应的方法被提出用于打破互易性,包括折射率调制[5,6]、光学非线性[7-9]、光子和声子系统中的角动量偏置[10-12],以及量子霍尔效应[13]。

此外,腔光力学领域主要研究电磁腔和机械振子之间通过辐射压力实现的相互作用[14-16]。基于光力相互作用,该领域取得了许多重要的进展,包括机械振子的基态冷却[17,18]、光力诱导透明[19-24]、微波放大[25-27]、光和微波光子之间的双向态转化[28-31]、声子激光[32-35]以及宇称-时间-对称性破缺的混沌[36]。最近,辐射压诱导的光力作用被用来破坏时间-反演对称性以及实现光[37-48]和微波频率[49-54]的电磁非互易性。这些工作多数依赖于控制场增强的光力耦合强度来实现非互易性。值得注意的是,基于光力系统的单向放大器可以通过库工程[45]、相干机械驱动[46]、蓝失谐的光泵浦[47,51]等方式实现。此外,单向放大器已经在超导微波电路中利用约瑟夫森非线性和参量泵浦实现[55-57]。

基于以上的成果,我们理论上研究了如何基于三腔光力系统实现量子极限的单向放大器,其中一个微波腔和两个直接耦合的光学腔分别耦合于一个共同的机械振子[49]。当微波腔被驱动至其红边带而两个光学腔分别被驱动至蓝边带时,微波和光学光子之间可以实现单向放大,主要包括两个增益过程和一个转化过程。我们发现放大的方向依赖于场增强的光力耦合强度之间的相对相位差,并且放大器的增益和带宽随着光力协同性的增强而变大。此外,机械噪声在协同性很大时可被抑制,并且这种相位保持的放大器可以达到量子极限。

与之前工作中的光学非互易性相比[37-48]，本节提出的方案可以实现光学光子和微波光子之间的单向放大，并且不要求光学腔和微波之间直接耦合。此外，另外一个光学腔被引入用来控制放大的方向。重要的是，提出的单向放大器能够抑制机械噪声的影响并且能达到增加噪声的量子极限值。最近，基于一个微波光力系统实现微波信号之间的量子极限的单向放大器方案被提出，其中两个微波腔分别耦合于两个机械振子[51]。与之相比，我们的工作可以实现三个腔之间的单向放大，主要包括了两个放大过程和一个转换过程，类似的过程已经在约瑟夫森参数转化器中实现[56]。此外，文献[51]中出现隔离时的相位差取决于泵浦场和腔场之间的失谐量，但是在我们的工作中当相位差为$\pm\pi/2$时即可出现单向放大，更加容易进行控制。

6.2.2 模型和理论

考虑如图 6.2.1 所示的光力系统，其中三个腔模 a_1、a_2、a_3 通过辐射压力分别耦合于一个共同的力学模式 b。此外，腔 a_2 和 a_3 之间直接耦合从而便于实现非互易性传输。腔 a_1 和腔 $a_2(a_3)$ 的共振频率可以截然不同，比如，腔 a_1 可以是微波腔而腔 a_2 和 a_3 是光学腔，反之亦然。这个模型在三个腔全部是光学腔时仍然成立，但是这里我们主要讨论如何基于该系统实现光学和微波光子之间的单向放大，这在目前的实验条件下是可以实现的[28,31,58-60]。我们注意到两个腔模耦合于一个共同的机械振子的物理系统已经在实验上实现。当腔 a_1 是微波腔而腔 a_2 和 a_3 是光学腔时，两个光学腔之间需要引入时间依赖的相互作用，可以通过将光学腔耦合到其他腔模或波导实现[49,60,61]。为了实现单向放大，我们对腔 a_1 施加红失谐的泵浦场而对腔 $a_2(a_3)$ 施加蓝失谐的泵浦场。该光力系统的哈密顿量为($\hbar=1$)

$$H = \sum_{k=1}^{3} \omega_k a_k^\dagger a_k + \omega_m b^\dagger b + \sum_{k=1}^{3} g_k a_k^\dagger a_k (b^\dagger + b) + J(a_2^\dagger a_3 + a_3^\dagger a_2) +$$
$$\sum_{k=1}^{3} \varepsilon_k (a_k^\dagger e^{-i\omega_{d,k} t} + a_k e^{i\omega_{d,k} t}) \tag{6.2.1}$$

其中，$a_k(a_k^\dagger)$ 是共振频率为 ω_k 的腔模 $a_k(k=1,2,3)$ 的湮灭(产生)算符，$b(b^\dagger)$ 是共振频率为 ω_m 的力学模式的湮灭(产生)算符。第三项表示腔模 a_k 与力学模式 b 之间的相互作用，其中 g_k 是单光子耦合强度。第四项表示

腔 a_2 和 a_3 之间的相互作用,其耦合强度为 J。最后一项表示泵浦场和各自的腔模之间的相互作用,其中 ε_k 和 $\omega_{d,k}$ 分别表示施加到腔模 a_k 的泵浦场的强度和频率。方程(6.2.1)中的驱动项通过幺正变换 $U = \exp\left(i\sum_{k=1}^{3}\omega_{d,k}a_k^+ a_k t\right)$ 可以变得不依赖于时间,新的哈密顿量 $H_{\rm rot} = UHU^+ - iU\partial U^+/\partial t$ 变为

$$H_{\rm rot} = \sum_{k=1}^{3}\Delta_k a_k^+ a_k + \omega_m b^+ b + \sum_{k=1}^{3} g_k a_k^+ a_k (b^+ + b) + J(a_2^+ a_3 + a_3^+ a_2) +$$

$$\sum_{k=1}^{3}\varepsilon_k (a_k^+ + a_k) \tag{6.2.2}$$

其中我们已经假定 $\omega_{d,2} = \omega_{d,3}$,$\Delta_k = \omega_k - \omega_{d,k}$ 是腔模 a_k 与各自的泵浦场之间的失谐量。

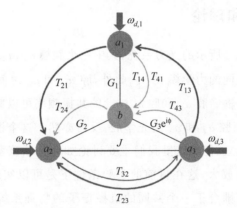

图 6.2.1　三腔光力系统示意图

(请扫 Ⅱ 页二维码看彩图)

根据方程(6.2.2),可以得到如下量子朗之万方程:

$$\dot{a}_1^+ = -(\kappa_1/2 - i\Delta)a_1^+ + ig_1 a_1^+(b^+ + b) + i\varepsilon_1 + \sqrt{\kappa_1}\, a_{1,\rm in}^+ \tag{6.2.3}$$

$$\dot{a}_2 = -(\kappa_2/2 + i\Delta_2)a_2 - ig_2 a_2(b^+ + b) - iJa_3 - i\varepsilon_2 + \sqrt{\kappa_2}\, a_{2,\rm in} \tag{6.2.4}$$

$$\dot{a}_3 = -(\kappa_3/2 + i\Delta_3)a_3 - ig_3 a_3(b^+ + b) - iJa_2 - i\varepsilon_3 + \sqrt{\kappa_3}\, a_{3,\rm in} \tag{6.2.5}$$

$$\dot{b}^+ = -(\gamma_m/2 - i\omega_m)b^+ + i\sum_k g_k a_k^+ a_k + \sqrt{\gamma_m}\, b_{\rm in}^+ \tag{6.2.6}$$

其中宏观地引入了衰减和噪声项。κ_1、κ_2、κ_3 分别是腔模 $a_k(k=1,2,3)$ 的衰减率,γ_m 是机械振子的衰减率。此外,$a_{k,\rm in}(k=1,2,3)$ 和 $b_{\rm in}$ 分别是腔模和力学模式的噪声算符,它们的平均值为零,并且满足如下关联函数:

$$\begin{cases} \langle a_{k,\text{in}}(t)a_{k,\text{in}}^+(t')\rangle = \delta(t-t'), \langle a_{k,\text{in}}^+(t)a_{k,\text{in}}(t')\rangle = 0 \\ \langle b_{\text{in}}(t)b_{\text{in}}^+(t')\rangle = (n_m+1)\delta(t-t'), \langle b_{\text{in}}^+(t)b_{\text{in}}(t')\rangle = n_m\delta(t-t') \end{cases} \quad (6.2.7)$$

其中我们假定热光子数为零,机械振子的热声子数 $n_m = 1/(e^{\hbar\omega_m/k_B T_m} - 1)$,这里 T_m 是环境的温度,k_B 是玻耳兹曼常数。

腔模和力学模式的稳态解可以通过令方程(6.2.3)~方程(6.2.6)为零并忽略所有的噪声项得到,满足以下的代数方程:

$$\alpha_1^* = \frac{i\varepsilon_1}{\kappa_1/2 - i\Delta_1'}, \quad \alpha_2 = -\frac{iJ\alpha_3 + i\varepsilon_2}{\kappa_2/2 + i\Delta_2'},$$

$$\alpha_3 = -\frac{iJ\alpha_2 + i\varepsilon_3}{\kappa_3/2 + i\Delta_3'}, \quad \beta^* = \frac{i\sum_k g_k |\alpha_k|^2}{\gamma_m/2 - i\omega_m} \quad (6.2.8)$$

其中,$\alpha_k(k=1,2,3)$ 和 β 分别表示腔模 a_k 和力学模式 b 的平均值,$\Delta_k' = \Delta_k + g_k(\beta^* + \beta)$ 是包含辐射压效应后有效的腔失谐量。接下来,我们将每个海森伯算符写成稳态解和小的涨落和的形式,即 $a_k = \alpha_k + \delta a_k, b = \beta + \delta b$,代入方程(6.2.3)~方程(6.2.6),从而可以得到线性化的量子朗之万方程:

$$\delta\dot{a}_1^+ = -(\kappa_1/2 - i\Delta_1')\delta a_1^+ + ig_1\alpha_1^*(\delta b^+ + \delta b) + \sqrt{\kappa_1} a_{1,\text{in}}^+ \quad (6.2.9)$$

$$\delta\dot{a}_2 = -(\kappa_2/2 + i\Delta_2')\delta a_2 - ig_2\alpha_2(\delta b^+ + \delta b) - iJ\delta a_3 + \sqrt{\kappa_2} a_{2,\text{in}} \quad (6.2.10)$$

$$\delta\dot{a}_3 = -(\kappa_3/2 + i\Delta_3')\delta a_3 - ig_3\alpha_3(\delta b^+ + \delta b) - iJ\delta a_2 + \sqrt{\kappa_3} a_{3,\text{in}} \quad (6.2.11)$$

$$\delta\dot{b}^+ = -(\gamma_m/2 - i\omega_m)\delta b^+ + i\sum_k g_k(\alpha_k^*\delta a_k + \alpha_k\delta a_k^+) + \sqrt{\gamma_m} b_{\text{in}}^+ \quad (6.2.12)$$

为了实现单向放大,调制泵浦场的频率满足 $\Delta_1' = \omega_m, \Delta_2' = \Delta_3' = -\omega_m$,即腔模 a_1 被驱动至红边带,而腔模 a_2 和 a_3 被驱动至蓝边带。在可分辨极限下,$\omega_m \gg \gamma_m, \kappa_k (k=1,2,3)$,可以作旋转波近似。为了简单起见,引入 $\delta a_k \to \delta a_k e^{-i\Delta_k' t}, \delta b \to \delta b e^{-i\omega_m t}, a_{k,\text{in}} \to a_{k,\text{in}} e^{-i\Delta_k' t}, b_{\text{in}} \to b_{\text{in}} e^{-i\omega_m t}$[62],从而变换到另一种相互作用框架,旋转波近似后方程(6.2.9)~方程(6.2.12)变成

$$\delta\dot{a}_1^+ = -\frac{\kappa_1}{2}\delta a_1^+ + iG_1 e^{i\phi_1}\delta b^+ + \sqrt{\kappa_1} a_{1,\text{in}}^+ \quad (6.2.13)$$

$$\delta\dot{a}_2 = -\frac{\kappa_2}{2}\delta a_2 - iG_2 e^{-i\phi_2}\delta b^+ - iJ\delta a_3 + \sqrt{\kappa_2} a_{2,\text{in}} \quad (6.2.14)$$

$$\delta\dot{a}_3 = -\frac{\kappa_3}{2}\delta a_3 - iG_3 e^{-i\phi_3}\delta b^+ - iJ\delta a_2 + \sqrt{\kappa_3} a_{3,\text{in}} \quad (6.2.15)$$

$$\delta\dot{b}^+ = -\frac{\gamma_m}{2}\delta b^+ + \mathrm{i}(G_1 \mathrm{e}^{-\mathrm{i}\phi_1}\delta a_1^+ + G_2 \mathrm{e}^{\mathrm{i}\phi_2}\delta a_2 + G_3 \mathrm{e}^{\mathrm{i}\phi_3}\delta a_3) + \sqrt{\gamma_m}b_{\mathrm{in}}^+ \quad (6.2.16)$$

其中我们假定 $g_k\alpha_k = g_k|\alpha_k|\mathrm{e}^{-\mathrm{i}\phi_k} = G_k \mathrm{e}^{-\mathrm{i}\phi_k}$，$G_k$ 为场增强的耦合强度。根据方程(6.2.13)~方程(6.2.16)，可得到线性化后的有效哈密顿量

$$H_{\mathrm{eff}} = G_1 \delta a_1 \delta b^+ \mathrm{e}^{\mathrm{i}\phi_1} + G_2 \delta a_2 \delta b \mathrm{e}^{\mathrm{i}\phi_2} + G_3 \delta a_2 \delta b \mathrm{e}^{\mathrm{i}\phi_3} + J\delta a_2 \delta a_3 + \mathrm{H.c.} \quad (6.2.17)$$

相位 $\phi_k(k=1,2,3)$ 可以通过重新定义算符 δa_k 和 δb 进行吸收，从而只有由 G_2、G_3 和 J 构成的闭环中的相位差 $\phi = \phi_3 - \phi_3$ 有物理意义。为了方便，可以令 $\phi_1 = \phi_2 = 0$，$\phi_3 = \phi$，并且忽略符号 δ，即 $\delta a_k \to a_k$ 和 $\delta b \to b$。因此，方程(6.2.13)~方程(6.2.16)可以写成矩阵形式：

$$\dot{\boldsymbol{\mu}} = \boldsymbol{M}\boldsymbol{\mu} + \sqrt{\boldsymbol{K}}\,\boldsymbol{\mu}_{\mathrm{in}} \quad (6.2.18)$$

其中矢量算符 $\boldsymbol{\mu} = (a_1^+, a_2, a_3, b^+)^{\mathrm{T}}$，$\boldsymbol{\mu}_{\mathrm{in}} = (a_{1,\mathrm{in}}^+, a_{2,\mathrm{in}}, a_{3,\mathrm{in}}, b_{\mathrm{in}}^+)^{\mathrm{T}}$，对角矩阵 $\boldsymbol{K} = \mathrm{Diag}[\kappa_1, \kappa_2, \kappa_3, \gamma_m]$，系数矩阵

$$\boldsymbol{M} = \begin{pmatrix} -\kappa_1/2 & 0 & 0 & \mathrm{i}G_1 \\ 0 & -\kappa_2/2 & -\mathrm{i}J & -\mathrm{i}G_2 \\ 0 & -\mathrm{i}J & -\kappa_3/2 & -\mathrm{i}G_3\mathrm{e}^{-\mathrm{i}\phi} \\ \mathrm{i}G_1 & \mathrm{i}G_2 & \mathrm{i}G_3\mathrm{e}^{\mathrm{i}\phi} & -\gamma_m/2 \end{pmatrix} \quad (6.2.19)$$

系统只有当矩阵 $\boldsymbol{M}$ 所有本征值的实部都为负数值时才是稳定的，其稳定性条件可以通过 Routh-Hurwitz 判据[63,64]解析地得到，但形式过于复杂，这里不再给出。但是，本工作中数值上检验了稳定性条件，并且确保所选参数满足稳定性条件。

通过引入算符的傅里叶变换

$$o(\omega) = \int_{-\infty}^{+\infty} o(t)\mathrm{e}^{\mathrm{i}\omega t}\mathrm{d}t \quad (6.2.20)$$

$$o^+(\omega) = \int_{-\infty}^{+\infty} o^+(t)\mathrm{e}^{\mathrm{i}\omega t}\mathrm{d}t \quad (6.2.21)$$

方程(6.2.18)在频域空间中的解为

$$\boldsymbol{\mu}(\omega) = -(\boldsymbol{M} + \mathrm{i}\omega\boldsymbol{I})^{-1}\sqrt{\boldsymbol{K}}\,\boldsymbol{\mu}_{\mathrm{in}}(\omega) \quad (6.2.22)$$

其中 $\boldsymbol{I}$ 表示单位矩阵。将方程(6.2.22)代入标准的输入-输出关系[65]$\boldsymbol{\mu}_{\mathrm{out}}(\omega) = \boldsymbol{\mu}_{\mathrm{in}}(\omega) - \sqrt{\boldsymbol{K}}\boldsymbol{\mu}(\omega)$，可以得到

第6章 基于三腔光力系统的单向放大器

$$\boldsymbol{\mu}_{\text{out}}(\omega) = \boldsymbol{T}(\omega)\boldsymbol{\mu}_{\text{in}}(\omega) \tag{6.2.23}$$

其中输出场算符矢量$\boldsymbol{\mu}_{\text{out}}(\omega)$是$\boldsymbol{\mu}_{\text{out}} = (a_{1,\text{out}}^1, a_{2,\text{out}}^+, a_{3,\text{out}}, b_{\text{out}}^+)^{\text{T}}$的傅里叶变换，并且传输矩阵

$$\boldsymbol{T}(\omega) = \boldsymbol{I} + \sqrt{\boldsymbol{K}}(\boldsymbol{M}+\mathrm{i}\omega\boldsymbol{I})^{-1}\sqrt{\boldsymbol{K}} \tag{6.2.24}$$

矩阵元$\boldsymbol{T}_{ij}(\omega)(i,j=1,2,3,4)$（分别对应于$a_1, a_2, a_3, b$）表示从模式$j$到模式$i$的传输幅度。

6.2.3 量子极限的单向放大器

本节详细讨论如何基于三腔光力系统实现单向放大器。我们已假定腔a_1被驱动至其红边带，而腔a_2和a_3被驱动至各自的蓝边带。为方便起见，首先讨论当输入场与腔频共振($\omega=0$)时腔a_1和a_2之间的单向放大。根据方程(6.2.19)和方程(6.2.24)，传输矩阵元

$$\boldsymbol{T}_{12}(\omega) = -\frac{\sqrt{\kappa_1\kappa_2}}{A(\omega)}G_1\Gamma_3(G_2+\mathrm{i}J\mathrm{e}^{\mathrm{i}\phi}G_3/\Gamma_3) \tag{6.2.25}$$

$$\boldsymbol{T}_{21}(\omega) = -\frac{\sqrt{\kappa_1\kappa_2}}{A(\omega)}G_1\Gamma_3(G_2+\mathrm{i}J\mathrm{e}^{-\mathrm{i}\phi}G_3/\Gamma_3) \tag{6.2.26}$$

式中

$$A(\omega) = \Gamma_1(\Gamma_2\Gamma_3\Gamma_m - \Gamma_2 G_3^2 - \Gamma_3 G_2^2 + \Gamma_m J^2 - 2\mathrm{i}J\cos\phi G_2 G_3) + G_1^2(\Gamma_2\Gamma_3 + J^2)$$

$$\Gamma_1 = -\kappa_1/2 + \mathrm{i}\omega$$
$$\Gamma_m = -\gamma_m/2 + \mathrm{i}\omega$$
$$\Gamma_2 = -\kappa_2/2 + \mathrm{i}\omega$$
$$\Gamma_3 = -\kappa_3/2 + \mathrm{i}\omega$$

不失一般性，考虑入射到腔a_1的探测场从腔a_2透射时可以被放大，而入射到腔a_2的探测场不能从腔a_1透射，即$|T_{21}|^2 > 1$，$|T_{12}|^2 = 0$。根据方程(6.2.25)和方程(6.2.26)很容易发现，当$\phi = -\pi/2$，$G_3 = G_2\kappa_3/(2J)$时$|T_{21}|^2 \neq 0$，$|T_{12}|^2 = 0$。接下来会发现该混杂系统中的相长干涉可以导致$|T_{21}|^2$大于1。此外，为了阻止输入场传输到其他模式而导致损失，要求$|T_{i1}/T_{21}| \ll 1 (i=3,4)$，通过调节耦合强度$J = \sqrt{\kappa_2\kappa_3}/2$可以满足这一条件。在满足上述条件的情况

下,共振时传输幅度

$$T_{21}(0) = -\frac{8\sqrt{\kappa_1\kappa_2}\,G_1 G_2}{4\kappa_2 G_1^2 - 4\kappa_1 G_2^2 + \kappa_1\kappa_2\gamma_m} = -\frac{2\sqrt{C_1 C_2}}{C_1 - C_2 + 1} \quad (6.2.27)$$

其中光力协同性 $C_k = 4G_k^2/(\kappa_k \gamma_m)$。此外,共振时传输矩阵

$$\boldsymbol{T}(0) = \begin{pmatrix} \dfrac{C_1 + C_2 - 1}{C_1 - C_2 + 1} & 0 & -\dfrac{2\mathrm{i}\sqrt{C_1 C_2}}{C_1 - C_2 + 1} & -\dfrac{2\mathrm{i}\sqrt{C_1}}{C_1 - C_2 + 1} \\ \dfrac{-2\sqrt{C_1 C_2}}{C_1 - C_2 + 1} & 0 & \mathrm{i}\dfrac{C_1 + C_2 + 1}{C_1 - C_2 + 1} & \dfrac{2\mathrm{i}\sqrt{C_2}}{C_1 - C_2 + 1} \\ 0 & \mathrm{i} & 0 & 0 \\ -\dfrac{2\mathrm{i}\sqrt{C_1}}{C_1 - C_2 + 1} & 0 & \dfrac{-2\sqrt{C_2}}{C_1 - C_2 + 1} & \dfrac{C_1 - C_2 - 1}{C_1 - C_2 + 1} \end{pmatrix} \quad (6.2.28)$$

接下来,将在数值上展示不同腔模之间的单向放大。参数主要根据最近的实验选取[17,18,20]:光学腔和微波腔的衰减率 $\kappa_k = 1 \sim 10$ MHz,场增强的耦合强度 G_k 可以达到几十兆赫兹。假定 $\kappa_1/(2\pi) = 2$ MHz, $\kappa_2/(2\pi) = \kappa_3/(2\pi) = 3$ MHz,机械振子的衰减率 $\gamma_m/(2\pi) = 30$ kHz。因为腔模 a_2 和 a_3 被驱动至蓝边带,可能导致系统不稳定。图 6.2.2 给出了稳定图随光力协同性 C_1 和 C_2 变化的情况,将从稳定区域中选出参数来研究单向放大器。从图中可以看出,当 $C_2 < C_1 \leqslant 60$ 时系统是稳定的,因此对于给定的 C_1 选取 $C_2 = C_1 - 0.1\sqrt{C_1}$ 确保系统工作在稳定区域。

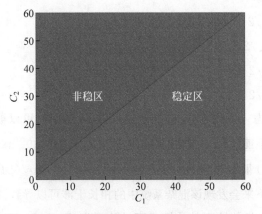

图 6.2.2 光力协同性 C_1 和 C_2 变化时的稳定图

(请扫Ⅱ页二维码看彩图)

根据方程(6.2.25)和方程(6.2.26),可以研究腔 a_1 和 a_2 之间的单向放大。图 6.2.3 给出了 $\phi = -\pi/2, 0, \pi/2$ 时 $|T_{12}|^2$,$|T_{21}|^2$ 随探测失谐量 ω 变化的情况。从图 6.2.3(a)可以看出,$|T_{21}(0)|^2 \approx 23$ dB,$|T_{21}|^2 = 0$,表示入射到腔 a_1 的共振探测场从腔 a_2 透射时可被放大很多,而从入射到腔 a_2 的探测场不能从腔 a_1 透射。因此,微波光子和光学光子之间的单向放大可以基于该光力系统实现。这一现象可以通过两个可能的传输路径之间的干涉效应进行解释,其中一条路径是 $a_1 \to b \to a_2$,透射强度正比于 G_2;另外一条路径是 $a_1 \to b \to a_3 \to a_2$,透射强度正比于 $iJe^{i\phi}G_3/\Gamma_3$。当 $\phi = -\pi/2, \omega = 0, G_3 = G_2\kappa_3/(2J)$ 时,两条路径之间的相长干涉导致从腔 a_1 到腔 a_2 的放大,但是相消干涉抑制了从腔 a_2 到腔 a_1 的传输。如果相位差 ϕ 调成 $\pi/2$,可以从图 6.2.3(c)看出,放大的方向与图 6.2.3(a)中的方向相反。在以上两种情况下,$\phi \neq n\pi$(n 是一个整数),时间-反演对称性被破坏,从而该系统中出现了非互易性传输现象。但如果 $\phi = \pm \pi$,图 6.2.3(b)和(d)表明探测失谐量 ω 变化时透射率 $|T_{21}|^2 = |T_{12}|^2$。不同方向的传输特性是互易的,并且没有单向放大。

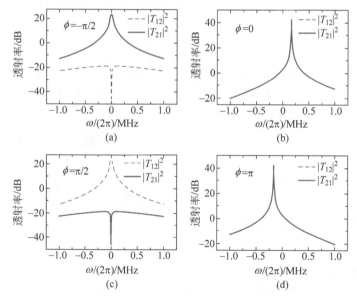

图 6.2.3 相位差 $\phi = -\pi/2, 0, \pi/2$ 时透射率 $|T_{12}|^2$,$|T_{21}|^2$ 随探测失谐量 ω 变化的曲线

(请扫 Ⅱ 页二维码看彩图)

图 6.2.4 讨论了当相位差 $\phi=-\pi/2$ 时三个腔模之间的单向放大情况。图 6.2.4(a)表明入射到腔 a_1 的信号场从腔 a_2 透射时可以被单向放大,图 6.2.4(b)表明入射到腔 a_2 上的信号场从腔 a_3 出射时透射率近似为 1,图 6.2.4(c)表明入射到腔 a_3 上的信号场从腔 a_1 透射时也可以被单向放大,但反过来则不行。因此,该光力系统中的单向放大包括两个增益过程和一个转化过程,这与基于约瑟夫森参量转化器实现的单向放大器情况类似[56]。此处三个腔可以看作单向放大器的三个端口,分别标记为信号输入(S)、闲置输入(I)和真空输入(V)端口。入射到 S 端口(放大器输入)的信号场从 I 端口(放大器输出)输出时被单向放大,并且入射到 V 端口的信号又近乎无损地透射回 S 端口。当相位差 $\phi=-\pi/2$,腔 a_3 可视为 S 端口,腔 a_1 为 I 端口,腔 a_2 为 V 端口。单向放大沿着路径 $a_3 \to a_1 \to a_2 \to a_3$,如图 6.2.4(d)所示。此外,如果相位差被调节成 $\phi=\pi/2$,放大的方向将变得相反,即沿着路径 $a_2 \to a_1 \to a_3 \to a_2$。在这种情况下,腔 a_2 为 S 端口,腔 a_3 为 V 端口。

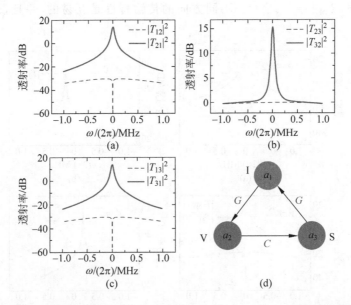

图 6.2.4 单向放大器。(a)~(c)透射率 $|T_{ij}|^2 (i,j=1,2,3)$ 随探测失谐量 $\omega/(2\pi)$ 变化的曲线;(d)放大过程的图形表示,其中 S、I 和 V 分别表示信号、闲置和真空端口,G 和 C 分别代表增益和转化过程

正如图 6.2.4 所示，当相位差 $\phi=-\pi/2$ 时，腔 a_3 作为放大器的输入端口而腔 a_1 作为放大器的输出端口。但是，包括力学模式在内的其他模式会不可避免地给放大器增加噪声。放大器增加的噪声数可以通过计算腔 a_1 的输出谱得到，由下式给出：

$$S_{1,\text{out}}(\omega) = \frac{1}{2}\int \frac{d\Omega}{2\pi} \langle a_{1,\text{out}}(\omega) a_{1,\text{out}}^+(\Omega) + a_{1,\text{out}}^+(\Omega) a_{1,\text{out}}(\omega) \rangle$$

$$= \frac{1}{2}|T_{11}(\omega)|^2 + \frac{1}{2}|T_{12}(\omega)|^2 + \frac{1}{2}|T_{13}(\omega)|^2 +$$

$$\left(n_m + \frac{1}{2}\right)|T_{14}(\omega)|^2 \tag{6.2.29}$$

其中利用了频域空间的噪声关联函数和关系式 $o^+(\omega)=[o(-\omega)]^+$。信号中增加的噪声定义为[51,66-68] $N_1(\omega) = G^{-1}\sum_{i\neq 3}(n_i+1/2)|T_{1i}(\omega)|^2$，其中单向放大器的功率增益 G 由下式给出：

$$G = |T_{31}(0)|^2 = \frac{4C_1 C_2}{C_1 - C_2 + 1} \tag{6.2.30}$$

共振时，腔 a_1 的输出中增加的噪声

$$N_1(0) = \frac{1}{2}\frac{(C_1+C_2-1)^2}{4C_1 C_2} + \left(n_m + \frac{1}{2}\right)\frac{1}{C_2} \tag{6.2.31}$$

其中已经假定腔模的热光子数 $n_1=n_2=n_3=0$，力学模式的热声子数 $n_4=n_m$。从方程(6.2.31)可以看出，通过增加光力协同性 C_2 可以抑制机械振子带来的热噪声影响。此外，$C_1 \geqslant C_2$ 且比较大时，我们发现零频时 $N_1(0) \to 1/2$，这是相位保持的线性放大器增加的噪声所能达到的量子极限[51,66-68]。

为了能够更加清楚地研究光力协同性对放大器的增益和增加的噪声带来的影响，我们在图 6.2.5 中示出了协同性 C_1 取不同值而 $C_2 = C_1 - 0.1\sqrt{C_1}$ 时(a)透射率 $|T_{13}|^2$ 和(b)增加的噪声 N_1 随探测失谐量 ω 变化的情况。图 6.2.5(a)表明，$|T_{13}|^2$ 的峰值和带宽随着 C_1 从 1 增大到 50 而不断变大。此外，图 6.2.5(b)表明，增加的量子数 N_1 随着 C_1 增大而减小。尤其是当 C_1 和 C_2 足够大时，共振处 N_1 可以达到极限值 1/2。因此，基于该三腔光力系统可以实现量子极限的单向放大器。

最后，图 6.2.5 表明该单向放大器的增益和带宽随光力协同性 C_1 增加

($C_2 = C_1 - 0.1\sqrt{C_1}$)时而变大。带宽和增益-带宽积的表达式可以通过以下方法近似得到。假定三个腔的衰减率相同,即$\kappa_1 = \kappa_2 = \kappa_3 = \kappa$,那么方程(6.2.12)的分母$A(\omega) \approx \frac{1}{8}\kappa^3\gamma_m(C_1 - C_2 + 1) - \frac{1}{4}\gamma_m\kappa^2(\kappa/\gamma_m + C_1 - 2C_2 + 2)i\omega$,这里只保留$\omega$的一阶项[51]。带宽$\Gamma$可以由满足$2|A(0)|^2 = |A(\omega)|^2$的最小的$|\omega|$值得到,近似表达式为$\Gamma = |\kappa(C_1 - C_2 + 1)/(\kappa/\gamma_m + C_1 - 2C_2 + 2)|$。当$\kappa/\gamma_m \gg \{1, C_1, C_2\}$时,带宽近似为$\Gamma \approx \gamma_m(C_1 - C_2 + 1)$,主要受到机械振子的衰减率限制。因此,增益-带宽积$P \equiv \Gamma\sqrt{G} \approx 2\gamma_m\sqrt{C_1 C_2}$。此外,隔离带宽也需要考虑,在带宽范围内可以得到足够的隔离率。在$\omega = 0$附近,反向的透射振幅$T_{12}(\omega) \approx -i\omega\sqrt{G}/\kappa$。因此,隔离带宽大概在$\kappa/\sqrt{G}$量级,可以通过增加腔衰减率$\kappa$进一步增宽。在本节讨论的放大器中,隔离带宽远大于增益带宽,因此信号场在增益带宽内可被单向放大。

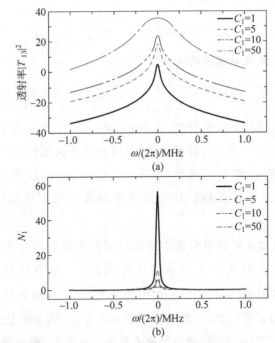

图 6.2.5 光力协同性$C_1 = \{1, 5, 10, 50\}, C_2 = C_1 - 0.1\sqrt{C_1}$时(a)透射率$|T_{13}|^2$和(b)腔$a_1$增加噪声$N_1$随探测失谐量$\omega/(2\pi)$变化的曲线

(请扫Ⅱ页二维码看彩图)

6.2.4 小结

本节提出了一种基于三腔光力系统实现量子极限的单向放大器的方案[69],其中一个微波腔被泵浦至红边带而两个直接耦合的光学腔被泵浦至各自的蓝边带。该系统中两个可能路径之间的量子干涉导致了微波和光学光子之间的单向放大,其中光力耦合强度之间的相位差起着非常重要的作用。此外,研究表明放大器的增益和带宽可以通过增加光力协同性而变大,在协同性比较大时该放大器增加的噪声可以达到量子极限。

参 考 文 献

[1] DEÁK L, FÜLÖP T. Reciprocity in quantum, electromagnetic and other wave scattering[J]. Ann. Phys., 2012, 327: 1050-1077.

[2] APLET L J, CARSON J W. A Faraday effect optical isolator[J]. Appl. Opt., 1964, 3: 544-545.

[3] POTTON R J. Reciprocity in optics[J]. Rep. Prog. Phys., 2004, 67: 717.

[4] BI L, HU J, JIANG P, et al. On-chip optical isolation in monolithically integrated non-reciprocal optical resonators[J]. Nat. Photon., 2011, 5: 758-762.

[5] YU Z, FAN S. Complete optical isolation created by indirect interband photonic transitions[J]. Nat. Photon., 2009, 3: 91-94.

[6] LIRA H, YU Z, FAN S, et al. Electrically driven nonreciprocity induced by interband photonic transition on a dilicon vhip[J]. Phys. Rev. Lett., 2012, 109: 033901.

[7] CHANG L, JIANG X, HUA S, et al. Parity-time symmetry and variable optical isolation in active-passive-coupled microresonators[J]. Nat. Photon., 2014, 8: 524-529.

[8] PENG B, ÖZDEMIR Ş K, LEI F, et al. Parity-time-symmetric whispering-gallery microcavities[J]. Nat. Phys., 2014, 10: 394-398.

[9] GUO X, ZOU C L, JUNG H, et al. On-chip strong coupling and efficient frequency conversion between telecom and visible optical modes[J]. Phys. Rev. Lett., 2016, 117: 123902.

[10] FLEURY R, SOUNAS D L, SIECK C F, et al. Sound isolation and giant linear nonreciprocity in a compact acoustic circulator[J]. Science, 2014, 343: 516-519.

[11] ESTEP N A, SOUNAS D L, SORIC J, et al. Magnetic-free non-reciprocity and isolation based on parametrically modulated coupled-resonator loops[J]. Nat.

Phys., 2014, 10: 923-927.

[12] WANG D W, ZHOU H T, GUO M J, et al. Optical diode made from a moving photonic crystal[J]. Phys. Rev. Lett., 2013, 110: 093901.

[13] MAHONEY A C, COLLESS J I, PAUKA S J, et al. On-chip microwave quantum Hall circulator[J]. Phys. Rev. X, 2017, 7: 011007.

[14] ASPELMEYER M, KIPPENBERG T J, MARQUARDT F. Cavity optomechanics [J]. Rev. Mod. Phys., 2014, 86: 1391.

[15] MARQUARDT F, GIRVIN S M. Optomechanics[J]. Physics, 2009, 2: 40.

[16] XIONG H, SI L G, LV X Y, et al. Review of cavity optomechanics in the weak-coupling regime: from linearization to intrinsic nonlinear interactions[J]. Sci. China Phys. Mech. Astron., 2015, 58: 1-13.

[17] CHAN J, ALEGRE T P M, SAFAVI-NAEINI A H, et al. Laser cooling of a nanomechanical oscillator into its quantum ground state[J]. Nature, 2011, 478: 89-92.

[18] TEUFEL J D, DONNER T, LI D, et al. Sideband cooling of micromechanical motion to the quantum ground state[J]. Nature, 2011, 475: 359-363.

[19] AGARWAL G S, HUANG S. Electromagnetically induced transparency in mechanical effects of light[J]. Phys. Rev. A, 2010, 81: 041803.

[20] WEIS S, RIVIÈRE R, DELÉGLISE S, et al. Optomechanically induced transparency[J]. Science, 2010, 330: 1520-1523.

[21] SAFAVI-NAEINI A H, ALEGRE T P M, CHAN J, et al. Electromagnetically induced transparency and slow light with optomechanics[J]. Nature, 2011, 472: 69-73.

[22] ZHOU X, HOCKE F, SCHLIESSER A, et al. Slowing, advancing and switching of microwave signals using circuit nanoelectromechanics[J]. Nat. Phys., 2013, 9: 179-184.

[23] JING H, ÖZDEMIR Ş K, GENG Z, et al. Optomechanically-induced transparency in parity-time-symmetric microresonators[J]. Sci. Rep., 2015, 5: 9663.

[24] JIAO Y, LÜ H, QIAN J, et al. Nonlinear optomechanics with gain and loss: amplifying higher-order sideband and group delay[J]. New J. Phys., 2016, 18: 083034.

[25] MASSEL F, HEIKKILÄ T T, PIRKKALAINEN J M, et al. Microwave amplification with nanomechanical resonators[J]. Nature, 2011, 480: 351-354.

[26] OCKELOEN-KORPPI C F, DAMSKÄGG E, PIRKKALAINEN J M, et al. Low-noise amplification and frequency conversion with a multiport microwave optomechanical device[J]. Phys. Rev. X, 2016, 6: 041024.

[27] TÓTH L D, BERNIER N R, NUNNENKAMP A, et al. A dissipative quantum reservoir for microwave light using a mechanical oscillator[J]. Nat. Phys., 2017, 13: 787-793.

[28] ANDREWS R W, PETERSON R W, PURDY T P, et al. Bidirectional and

efficient conversion between microwave and optical light[J]. Nat. Phys., 2014, 10: 321-326.

[29] TIAN L. Optoelectromechanical transducer: reversible conversion between microwave and optical photons[J]. Ann. Phys., 2015, 527: 1.

[30] DONG C, FIORE V, KUZYK M C, et al. Optical wavelength conversion via optomechanical coupling in a silica resonator[J]. Ann. Phys., 2015, 527: 100.

[31] HILL J T, SAFAVI-NAEINI A H, CHAN J, et al. Coherent optical wavelength conversion via cavity optomechanics[J]. Nat. Commun., 2012, 3: 1196.

[32] GRUDININ I S, LEE H, PAINTER O, et al. Phonon laser action in a tunable two-level system[J]. Phys. Rev. Lett., 2010, 104: 083901.

[33] COHEN J D, MEENEHAN S M, MACCABE G S, et al. Phonon counting and intensity interferometry of a nanomechanical resonator[J]. Nature, 2015, 520: 522-525.

[34] JING H, ÖZDEMIR S K, LÜ X Y, et al. PT-symmetric phonon laser[J]. Phys. Rev. Lett., 2014, 113: 053604.

[35] LÜ H, ÖZDEMIR S K, KUANG L M, et al. Exceptional points in random-defect phonon lasers[J]. Phys. Rev. Applied, 2017, 8: 044020.

[36] LÜ X Y, JING H, MA J Y, et al. PT-symmetry-breaking chaos in optomechanics[J]. Phys. Rev. Lett., 2015, 114: 253601.

[37] HAFEZI M, RABL P. Optomechanically induced non-reciprocity in microring resonators[J]. Opt. Express, 2012, 20: 7672-7684.

[38] METELMANN A, CLERK A A. Nonreciprocal photon transmission and amplification via reservoir engineering[J]. Phys. Rev. X, 2015, 5: 021025.

[39] KIM J, KUZYK M C, HAN K, et al. Non-reciprocal Brillouin scattering induced transparency[J]. Nat. Phys., 2015, 11: 275.

[40] DONG C H, SHEN Z, ZOU C L, et al. Brillouin-scattering-induced transparency and non-reciprocal light storage[J]. Nat. Commun., 2015, 6: 6193.

[41] SHEN Z, ZHANG Y L, CHEN Y, et al. Experimental realization of optomechanically induced non-reciprocity[J]. Nat. Photon., 2016, 10: 657.

[42] XU X W, LI Y. Optical non-reciprocity and optomechanical circulator in three-mode optomechanical systems[J]. Phys. Rev. A, 2015, 91: 053854.

[43] RUESINK F, MIRI M A, ALÙ A, et al. Nonreciprocity and magnetic-free isolation based on optomechanical interactions [J]. Nat. Commun., 2016, 7: 13662.

[44] MIRI M A, RUESINK F, VERHAGEN E, et al. Optical nonreciprocity based on optomechanical coupling[J]. Phys. Rev. Appl., 2017, 7: 064014.

[45] FANG K, LUO J, METELMANN A, et al. Generalized non-reciprocity in an optomechanical circuit via synthetic magnetism and reservoir engineering[J]. Nat. Phys., 2017, 13: 465-471.

[46] LI Y, HUANG Y Y, ZHANG X Z, et al. Optical directional amplification in a

three-mode optomechanical system[J]. Opt. Express, 2017, 25: 18907-18916.

[47] SHEN Z, ZHANG Y L, CHEN Y, et al. Reconfigurable optomechanical circulator and directional amplifier[J]. Nat. Commun., 2018, 9: 1797.

[48] SOHN D B, KIM S, BAHL G. Time-reversal symmetry breaking with acoustic pumping of nanophotonic circuits[J]. Nat. Photon., 2018, 12: 91-97.

[49] TIAN L, LI Z. Nonreciprocal quantum-state conversion between microwave and optical photons[J]. Phys. Rev. A, 2017, 96: 013808.

[50] XU X W, LI Y, CHEN A X, et al. Nonreciprocal conversion between microwave and optical photons in electro-optomechanical systems[J]. Phys. Rev. A, 2016, 93: 023827.

[51] MALZ D, TÓTH L D, BERNIER N R, et al. Quantum-limited directional amplifiers with optomechanics[J]. Phys. Rev. Lett., 2018, 120: 023601.

[52] PETERSON G A, LECOCQ F, CICAK K, et al. Demonstration of efficient nonreciprocity in a microwave optomechanical circuit[J]. Phys. Rev. X, 2017, 7: 031001.

[53] BERNIER N R, TÓTH L D, KOOTTANDAVIDA A, et al. Nonreciprocal reconfigurable microwave optomechanical circuit [J]. Nat. Commun., 2017, 8: 604.

[54] BARZANJEH S, WULF M, PERUZZO M, et al. Mechanical on-chip microwave circulator[J]. Nat. Commun., 2017, 8: 953.

[55] ABDO B, SLIWA K, FRUNZIO L, et al. Directional amplification with a Josephson circuit[J]. Phys. Rev. X, 2013, 3: 031001.

[56] SLIWA K M, HATRIDGE M, NARLA A, et al. Reconfigurable Josephson circulator/directional amplifier[J]. Phys. Rev. X, 2015, 5: 041020.

[57] LECOCQ F, RANZANI L, PETERSON G A, et al. Nonreciprocal microwave signal processing with a field-programmable Josephson amplifier[J]. Phys. Rev. Appl., 2017, 7: 024028.

[58] BOCHMANN J, VAINSENCHER A, AWSCHALOM D D, et al. Nanomechanical coupling between microwave and optical photons[J]. Nat. Phys., 2013, 9: 712-716.

[59] PALOMAKI T A, TEUFEL J D, SIMMONDS R W, et al. Entangling mechanical motion with microwave fields[J]. Science, 2013, 342: 710-713.

[60] SATO Y, TANAKA Y, UPHAM J, et al. Strong coupling between distant photonic nanocavities and its dynamic control[J]. Nat. Photon., 2012, 6: 56-61.

[61] FANG K, YU Z, FAN S. Realizing effective magnetic field for photons by controlling the phase of dynamic modulation[J]. Nat. Photon., 2012, 6: 782-787.

[62] GENES C, MARI A, TOMBESI P, et al. Robust entanglement of a micromechanical resonator with output optical fields[J]. Phys. Rev. A, 2008, 78: 032316.

[63] DEJESUS E X, KAUFMAN C. Routh-Hurwitz criterion in the examination of eigenvalues of a system of nonlinear ordinary differential equations[J]. Phys. Rev.

A, 1987, 35: 5288.
[64] GRADSHTEYN I S, RYZHIK I M. Table of Integrals[M]. Series and Products: Academic Press, 1980.
[65] GARDINER C W, COLLETT M J. Input and output in damped quantum system: quantum stochastic differential equations and the master equation[J]. Phys. Rev. A, 1985, 31: 3761.
[66] METELMANN A, CLERK A A. Quantum-limited amplification via reservoir engineering[J]. Phys. Rev. Lett., 2014, 112: 133904.
[67] CAVES C M. Quantum limits on noise in linear amplifiers[J]. Phys. Rev. D, 1982, 26: 1817.
[68] CLERK A A, DEVORET M H, GIRVIN S M, et al. Introduction to quantum noise, measurement, and amplification[J]. Rev. Mod. Phys., 2010, 82: 1155.
[69] JIANG C, JI B W, CUI Y S, et al. Quantum-limited directional amplifier based on a triple-cavity optomechanical system[J]. Opt. Express, 2018, 26: 15255-15267.

6.3 微波和光学光子之间相位敏感的单向放大器

6.3.1 引言

线性放大器可以放大弱的输入信号，在信号处理和通信系统中有着广泛的应用。角频率为 ω 的与时间有关的输入信号可以分解成形式为 $U\cos\omega t + V\sin\omega t$ 的两个正交分量，其中 U 和 V 分别是两个正交分量的振幅。根据输入信号的两个正交分量的增益，线性放大器可以分为两种[1]：一种是相位不敏感的放大器，其中信号的两个正交分量被放大相同的倍数 $\sqrt{G}$ ($G>1$)，并且对易关系导致最小的增加噪声数是 1/2；另外一种是相位敏感的放大器，其中两个正交分量被放大不同的倍数 $\sqrt{G_U}$ 和 $\sqrt{G_V}$。如果其中一个正交分量被放大 $\sqrt{G}$ 倍而另外一个分量被抑制 $1/\sqrt{G}$ 倍，则相位敏感的放大器可以是无噪声的。

此外，如果输入信号沿着一个方向被放大而在相反的方向被抑制，那么放大器就是单向的。包括单向放大器、隔离器和循环器在内的非互易性器件是经典和量子光子电路中非常重要的部分[2]。实现非互易性最基本的要求是打破时间-反演-对称性[3]。在过去几十年中，多种不同的方法被用来实现非互易性，比如磁光效应[4,5]、动态调制[6,7]、光学非线性[8-10]、角动量偏置[11-13]

等。此外,不同泵浦条件下超导微波电路中的约瑟夫森非线性被用来实现循环器和单向放大器[14,15]。最近,梅特尔曼和科拉克提出了一种基于库工程方法实现非互易性传输和放大的方案[16],之后他们还进一步将该方法推广为自反馈方法中的一种[17]。

另外,腔光力学领域在过去十几年中取得了非常重要的进展[18-20],这一领域主要研究腔和力学模式之间通过辐射压形成的相互作用。通过对腔施加一束强的泵浦场,有效光力耦合强度可以被大大提高,并且系统的光学响应受到机械运动的影响也会发生改变。如果腔场受到红边带的泵浦场驱动,探测场和产生的反斯托克斯场之间的相消干涉会导致光力诱导透明现象的出现[21-25]。此外,当腔场受到蓝边带的泵浦场驱动时,探测场和产生的斯托克斯场之间的相长干涉可以引起光力诱导吸收和放大现象[26-30]。最近,兰扎尼(Ranzani)和奥门多(Aumentado)的研究表明在耦合模系统中多路径干涉和辅助模式中的损耗可以导致非互易性[31]。不同路径之间的相长干涉使得光可以沿着某一方向传输,但是相消干涉抑制了相反方向的传输。在耦合光力系统中,光力诱导透明和放大被拓展到研究非互易性传输和单向放大[32-54]。尤其是,马尔兹(Malz)等提出了一种基于由两个微波腔和两个机械振子构成的微波光力系统实现微波信号的相位敏感的单向放大的方案[46],其中有一个腔同时被驱动至红边带和蓝边带。这样的一种双色驱动方案最近已被用来实现机械振子的压缩和相位敏感的放大[55-59]。

本节我们提出了一种基于多模光力系统实现微波和光学光子之间实现相位灵敏的单向放大方案,该光力系统由一个微波腔和两个直接耦合的光学腔分别耦合于一个共同的机械振子形成[38]。我们发现,当微波腔被同时驱动在红边带和蓝边带而光学腔只被驱动在红边带时,入射在微波(光学)腔而从光学(微波)腔出射的信号的正交分量之间可以实现相位敏感的单向放大现象。放大的现象可以通过有效光力耦合强度之间的相位差进行控制。此外,研究表明,本节提出的单向放大器的增益-带宽积不受限制,并且增加的噪声数可以达到量子极限值。

6.3.2 模型和理论

考虑如图 6.3.1 所示的光力系统,其中三个腔模 $a_k(k=1,2,3)$ 通过辐射

第 6 章 基于三腔光力系统的单向放大器

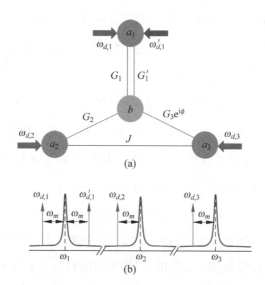

图 6.3.1 (a)三腔光力系统示意图。腔模 a_1、a_2、a_3 通过辐射压分别耦合于一个共同的机械振子 b,有效光力耦合强度分别为 G_1、G_2、$G_3 e^{i\phi}$。腔 a_2 和 a_3 之间的耦合强度为 J。腔 a_1 同时被驱动至红边带和蓝边带,泵浦场频率分别为 $\omega_{d,1}$ 和 $\omega'_{d,1}$,但腔 a_2 和 a_3 只受到红边带的泵浦场驱动,泵浦场频率分别为 $\omega_{d,2}$ 和 $\omega_{d,3}$。(b)腔模和泵浦场频率关系示意图
(请扫 II 页二维码看彩图)

压分别耦合于一个共同的机械振子 b,并且腔模 a_2 和 a_3 之间存在直接耦合,耦合强度为 J。这个光力系统的哈密顿量可写为

$$H = \sum_{k=1}^{3} \omega_k a_k^+ a_k + \omega_m b^+ b + \sum_{k=1}^{3} g_k a_k^+ a_k (b^+ + b) + J(a_2^+ a_3 + a_2 a_3^+) + \sum_{k=1}^{3} i\varepsilon_k (a_k^+ e^{-i\omega_{d,k} t} - a_k e^{i\omega_{d,k} t}) + i\varepsilon'_1 (a_1^+ e^{-i\omega_{d,1} t} - a_1 e^{i\omega_{d,1} t}) \quad (6.3.1)$$

其中,ω_k 和 ω_m 分别是腔模 a_k($k=1,2,3$)和机械振子 b 的共振频率,且腔模 a_k 和机械振子 b 之间的单光子耦合强度为 g_k。第四项表示腔 a_2 和 a_3 之间的相互作用,耦合强度为 J。最后两项表示驱动场与各自腔模之间的相互作用,其中 $\varepsilon_k(\omega_{d,k})$ 是红失谐的驱动场的振幅(频率),而 $\varepsilon'_1(\omega'_{d,1})$ 是施加到腔模 a_1 的蓝失谐的驱动场的振幅(频率)。通过将每个腔模写成稳态值加上涨落的和的形式,即 $a_1 \to \alpha_1 e^{-i\omega_{d,1} t} + \alpha'_1 e^{-i\omega'_{d,1} t} + a_1$,$a_k \to \alpha_k e^{-i\omega_{d,k} t} + a_k$($k=2,3$),方程(6.3.1)可以被线性化。在关于 $\sum_{k=1}^{3} \omega_k a_k^+ a_k + \omega_m b^+ b$ 的旋转框架下忽略掉

旋转频率为 $\pm 2\omega_m$ 的项之后，线性化的哈密顿量可表示为

$$H = G_1 a_1 b^+ \mathrm{e}^{\mathrm{i}\phi_1} + G_2 a_2 b^+ \mathrm{e}^{\mathrm{i}\phi_2} + G_3 a_3 b^+ \mathrm{e}^{\mathrm{i}\phi_3} + G_1' a_1 b \mathrm{e}^{\mathrm{i}\phi_1'} + J a_2 a_3^+ + \mathrm{H.c.} \tag{6.3.2}$$

其中我们已假定 $\omega_{d,k} - \omega_k = -\omega_m$（红边带）和 $\omega_{d,1}' - \omega_1 = \omega_m$（蓝边带）；$\omega_2 = \omega_3$；$g_k \alpha_k = G_k \mathrm{e}^{-\mathrm{i}\phi_k}$ 和 $g_1 \alpha_1' = G_1' \mathrm{e}^{-\mathrm{i}\phi_1'}$，其中 G_k 和 G_1' 是有效光力耦合强度。相位 $\phi_k (k=1,2,3)$ 和 ϕ_1' 可以通过重新定义算符 a_k 和 b 而被吸收掉，只有相位差 $\phi = \phi_3 - \phi_2$ 有物理效应。因此，方程(6.3.2)变为

$$H = G_1 a_1 b^+ + G_2 a_2 b^+ + G_3 a_3 b^+ \mathrm{e}^{\mathrm{i}\phi} + G_1' a_1 b + J a_2 a_3^+ + \mathrm{H.c.} \tag{6.3.3}$$

方程(6.3.3)中，诸如 $G_1 a_1 b^+ + \mathrm{H.c.}$ 之类的"分束器"相互作用可以导致光力诱导透明和量子态转化，而"双模压缩"相互作用 $G_1' a_1 b + \mathrm{H.c.}$ 可以产生光力诱导吸收和放大[26]，并且对腔模 a_1 的双色驱动可以用来实现相位敏感的放大[46]。此外，通过将腔 a_2 和 a_3 直接耦合，该光力系统形成了一个封闭的相互作用环 $b \to a_3 \to a_2 \to b$，这样就可以实现多路径干涉。因此，该系统中可以实现相位敏感的单向放大。定义正交分量

$$U_i = \frac{a_i + a_i^+}{\sqrt{2}}, \quad V_i = \frac{a_i - a_i^+}{\sqrt{2\mathrm{i}}}, \quad X = \frac{b + b^+}{\sqrt{2}}, \quad P = \frac{b - b^+}{\sqrt{2\mathrm{i}}} \tag{6.3.4}$$

满足对易关系 $[U_i, V_i] = \mathrm{i}$，$[X, P] = \mathrm{i}$，方程(6.3.3)可以改写为

$$H = 2G_1 U_1 X + G_2 (U_2 X + V_2 P) + G_3 [(U_3 X + V_3 P) \cos\phi +$$
$$(U_3 P - V_3 X) \sin\phi + J(U_2 U_3 + V_2 V_3)] \tag{6.3.5}$$

其中假定了 $G_1 = G_1'$（腔 a_1 的两个驱动场的强度相等）。因为 U_1 与哈密顿量(6.3.5)对易，因此它是一个量子非破坏性观测量。

根据方程(6.3.5)，可以得到如下量子朗之万方程：

$$\dot{U}_1 = -\frac{\kappa_1}{2} U_1 + \sqrt{\kappa_1} U_{1,\mathrm{in}} \tag{6.3.6}$$

$$\dot{V}_1 = -2G_1 X - \frac{\kappa_1}{2} V_1 + \sqrt{\kappa_1} V_{1,\mathrm{in}} \tag{6.3.7}$$

$$\dot{U}_2 = -\frac{\kappa_2}{2} U_2 + J V_3 + G_2 P + \sqrt{\kappa_2} U_{2,\mathrm{in}} \tag{6.3.8}$$

$$\dot{V}_2 = -\frac{\kappa_2}{2}V_2 - JU_3 - G_2 X + \sqrt{\kappa_2}V_{2,\text{in}} \tag{6.3.9}$$

$$\dot{U}_3 = JV_2 - \frac{\kappa_3}{2}U_3 - G_3\sin\phi X + G_3\cos\phi P + \sqrt{\kappa_3}U_{3,\text{in}} \tag{6.3.10}$$

$$\dot{V}_3 = -JU_2 - \frac{\kappa_3}{2}V_3 - G_3\cos\phi X - G_3\cos\phi P + \sqrt{\kappa_3}V_{3,\text{in}} \tag{6.3.11}$$

$$\dot{X} = G_2 V_2 + G_3\sin\phi U_3 + G_3\cos\phi V_3 - \frac{\gamma_m}{2}X + \sqrt{\gamma_m}X_{\text{in}} \tag{6.3.12}$$

$$\dot{P} = -2G_1 U_1 - G_2 U_2 - G_3\cos\phi U_3 + G_3\sin\phi V_3 - \frac{\gamma_m}{2}P + \sqrt{\gamma_m}P_{\text{in}} \tag{6.3.13}$$

其中，κ_k 是腔模 $a_k(k=1,2,3)$ 的衰减率而 γ_m 是机械振子 b 的衰减率。$U_{k,\text{in}}(V_{k,\text{in}})$ 和 $X_{\text{in}}(P_{\text{in}})$ 分别是腔模 a_k 和力学模式 b 的噪声算符，其平均值为零。将方程(6.3.6)～方程(6.3.13)变换到频域空间并消除力学模式，可以得到如下形式的矩阵方程：

$$\boldsymbol{T}(\omega)\boldsymbol{A}(\omega) + \boldsymbol{M}(\omega)\boldsymbol{B}_{\text{in}}(\omega) = \boldsymbol{L}\boldsymbol{A}_{\text{in}}(\omega) \tag{6.3.14}$$

其中，矢量

$$\boldsymbol{A}(\omega) = [U_1(\omega), V_1(\omega), U_2(\omega), V_2(\omega), U_3(\omega), V_3(\omega)]^{\text{T}}$$

$$\boldsymbol{A}_{\text{in}}(\omega) = [U_{1,\text{in}}(\omega), V_{1,\text{in}}(\omega), U_{2,\text{in}}(\omega), V_{2,\text{in}}(\omega), U_{3,\text{in}}(\omega), V_{3,\text{in}}(\omega)]^{\text{T}}$$

$$\boldsymbol{B}_{\text{in}}(\omega) = [X_{\text{in}}(\omega), P_{\text{in}}(\omega)]^{\text{T}}$$

系数矩阵

$$\boldsymbol{L} = \text{Diag}(\sqrt{\kappa_1}, \sqrt{\kappa_1}, \sqrt{\kappa_2}, \sqrt{\kappa_2}, \sqrt{\kappa_3}, \sqrt{\kappa_3}) \tag{6.3.15}$$

$$\boldsymbol{M} = \begin{pmatrix} 0 & 0 \\ 2G_1\chi_m\sqrt{\gamma_m} & 0 \\ 0 & -G_2\chi_m\sqrt{\gamma_m} \\ G_2\chi_m\sqrt{\gamma_m} & 0 \\ G_3\sin\phi\chi_m\sqrt{\gamma_m} & -G_3\cos\phi\chi_m\sqrt{\gamma_m} \\ G_3\cos\phi\chi_m\sqrt{\gamma_m} & G_3\sin\phi\chi_m\sqrt{\gamma_m} \end{pmatrix} \tag{6.3.16}$$

$$T = \begin{pmatrix} \chi_1^{-1} & 0 & 0 & 0 & 0 & 0 \\ 0 & \chi_1^{-1} & 0 & 2G_1G_2\chi_m & 2A_2 & 2A_1 \\ 2G_1G_2\chi_m & 0 & \chi_2^{-1}+G_2^2\chi_m & 0 & B_1 & -(B_2+J) \\ 0 & 0 & 0 & \chi_2^{-1}+G_2^2\chi_m & B_2+J & B_1 \\ 2A_1 & 0 & B_1 & B_2-J & \chi_3^{-1}+G_3^2\chi_m & 0 \\ -2A_2 & 0 & J-B_2 & B_1 & 0 & \chi_3^{-1}+G_3^2\chi_m \end{pmatrix}$$

(6.3.17)

这里定义 $\chi_k(\omega) = (\kappa_k/2 - \mathrm{i}\omega)^{-1}$,$\chi_m(\omega) = (\gamma_m/2 - \mathrm{i}\omega)^{-1}$,$A_1 = G_1G_3\cos\phi\chi_m$,$A_2 = G_1G_3\sin\phi\chi_m$,$B_1 = G_2G_3\cos\phi\chi_m$,$B_2 = G_2G_3\sin\phi\chi_m$。根据方程(6.3.14)和输入-输出关系[60] $A_{\mathrm{out}}(\omega) = A_{\mathrm{in}}(\omega) - LA(\omega)$,可以得到 $A_{\mathrm{out}}(\omega) = S(\omega)A_{\mathrm{in}}(\omega) + LT^{-1}(\omega)M(\omega)B_{\mathrm{in}}(\omega)$,其中散射矩阵 $S(\omega) = I_{6\times 6} - LT^{-1}(\omega)L$。矩阵元 $S_{ij} \equiv S_{j\to i}(i,j = U_1,V_1,U_2,V_2,U_3,V_3)$ 代表从正交分量 j 散射到分量 i 时对应的振幅。

基于散射矩阵 S,我们可以研究与腔频共振的输入信号的相位敏感的单向放大。在文献[31]中,兰扎尼和奥门多证实非互易性由多路径干涉引起,并且完全非互易性出现时回路相位应为 $\pm\pi/2$。他们的结果在本节研究的光力系统中依然有效。两个不同腔模之间的传输存在着两个不同的路径。比如考虑腔模 a_1 和 a_2 时,一条路径沿着 $a_1 \to b \to a_2$,而另一条路径沿着 $a_1 \to b \to a_3 \to a_2$。两条路径之间的干涉效应可以引起单向放大,其中不同路径之间的相位差(即 $b \to a_3 \to a_2 \to b$ 回路中的相位和)$\phi = \phi_3 - \phi_2$ 应该为 $\pm\pi/2$。如果相位差 ϕ 等于 $-\pi/2$ 时,可以得到

$$S_{U_1 \to U_2} = \frac{16G_1\sqrt{\kappa_2}(G_2\kappa_3 + 2G_3J)}{\sqrt{\kappa_1}(4\gamma_mJ^2 + 4\kappa_2G_3^2 + \gamma_m\kappa_2\kappa_3)} \quad (6.3.18)$$

$$S_{V_2 \to V_1} = \frac{16G_1\sqrt{\kappa_2}(G_2\kappa_3 - 2G_3J)}{\sqrt{\kappa_1}(4\gamma_mJ^2 + 4\kappa_2G_3^2 + 4\kappa_3G_2^2 + \gamma_m\kappa_2\kappa_3)} \quad (6.3.19)$$

$$S_{U_1 \to V_3} = \frac{16G_1\sqrt{\kappa_3}(G_2\kappa_3 - 2G_3J)}{\sqrt{\kappa_1}(4\gamma_mJ^2 + 4\kappa_2G_3^2 + 4\kappa_3G_2^2 + \gamma_m\kappa_2\kappa_3)} \quad (6.3.20)$$

$$S_{U_1 \to V_2} = S_{U_2 \to U_1} = S_{V_1 \to V_2} = S_{U_1 \to U_3} = 0 \quad (6.3.21)$$

考虑腔模 a_1 和 a_2 之间的非互易性传输时,可以发现如果 $G_3 = G_2\kappa_3/(2J)$,

则 $S_{U_1 \to U_2} > 0$，且 $S_{U_2 \to U_1} = S_{V_1 \to V_2} = S_{V_2 \to V_1} = 0$。此外，为了避免 U_1 散射到从腔 a_3 输出的信号中，可以进一步调节耦合强度 $J = G_3 \kappa_2/(2G_2)$ 使得 $S_{U_1 \to V_3} = 0$。因此，非互易性条件包括

$$\phi = -\pi/2, \quad J = \sqrt{\kappa_2 \kappa_3}/2, \quad G_3 = G_2 \kappa_3/(2J) \tag{6.3.22}$$

在上述条件下，可以得到以下重要的结果：

$$U_{1,\text{out}} = -U_{1,\text{in}} \tag{6.3.23}$$

$$V_{1,\text{out}} = -V_{1,\text{in}} - \frac{4\sqrt{C_1 C_2}}{C_2 + 1} U_{3,\text{in}} + \frac{4\sqrt{C_1}}{C_2 + 1} X_{\text{in}} \tag{6.3.24}$$

$$U_{2,\text{out}} = \frac{4\sqrt{C_1 C_2}}{C_2 + 1} U_{1,\text{in}} + \frac{C_2 - 1}{C_2 + 1} V_{3,\text{in}} - \frac{4\sqrt{C_2}}{C_2 + 1} P_{\text{in}} \tag{6.3.25}$$

$$V_{2,\text{out}} = -\frac{C_2 - 1}{C_2 + 1} U_{3,\text{in}} + \frac{2\sqrt{C_2}}{C_2 + 1} X_{\text{in}} \tag{6.3.26}$$

$$U_{3,\text{out}} = -V_{2,\text{in}} \tag{6.3.27}$$

$$V_{3,\text{out}} = U_{2,\text{in}} \tag{6.3.28}$$

其中光力协同性 $C_k = 4G_k^2/(\kappa_k \gamma_m)$。方程(6.3.25)表明入射到腔模 a_1 的信号的 U 分量从腔模 a_2 输出时可以被放大，但是方程(6.3.23)和方程(6.3.25)表明，入射到腔模 a_2 的信号不能从腔模 a_1 输出。因此，这里提出的相位敏感的放大器是单向的，且放大器的增益为

$$G = |S_{U_1 \to U_2}(0)|^2 = \frac{16 C_1 C_2}{(C_2 + 1)^2} \tag{6.3.29}$$

此外，当我们研究从 U_1 分量到 U_2 分量的单向放大时，除了 U_1 之外的其他分量到 U_2 分量的散射应该被当作噪声。增加的噪声可以通过[1,46,61] $N_{\text{add}}(\omega) = G^{-1} \sum_{k \neq U_1} (n_k + 1/2) |S_{k \to U_2}(\omega)|^2$ 计算得到，其中 n_k 是腔模 a_k 的热光子数，而 n_m 是力学模式的热声子数。根据方程(6.3.25)，可以得到

$$N_{2,\text{add}} = \frac{(C_2 - 1)^2}{16 C_1 C_2}\left(n_3 + \frac{1}{2}\right) + \frac{1}{4 C_1}\left(n_m + \frac{1}{2}\right) \tag{6.3.30}$$

如果选择 $C_1 = C_2^2 \gg 1$，则增加的噪声 $N_{2,\text{add}} \to 0$。因此，这里提出的相位敏感的单向放大器可以达到零噪声的量子极限值。

6.3.3 相位敏感的单向放大器

本节我们将基于上述讨论展示关于相位敏感的单向放大器的数值结果。

用到的参数根据最近的工作选取[38,49],具体如下:$\kappa_1/(2\pi)=2\,\text{MHz}$,$\kappa_2/(2\pi)=3\,\text{MHz}$,$\kappa_3/(2\pi)=3\,\text{MHz}$,$\gamma_m/(2\pi)=30\,\text{kHz}$。不失一般性,我们首先主要讨论腔模 a_1 和 a_2 分量之间的单向放大现象。

图 6.3.2 给出了相位差 ϕ 分别等于 $-\pi/2$ 和 $\pi/2$ 时散射几率 $|S_{U_1\to U_2}|^2$ 和 $|S_{V_2\to V_1}|^2$ 随探测失谐量 ω 变化的情况。从图 6.3.2(a) 可以看出,当探测场与腔频共振时($\omega=0$),$|S_{U_1\to U_2}|^2\approx 20\,\text{dB}$(根据 $10\log_{10}|S_{U_1\to U_2}|^2$ 换算为以 dB 为单位)且 $|S_{V_2\to V_1}|^2=0$。此外,我们之前已经得到当探测频率 ω 变化时,$S_{U_2\to U_1}$ 和 $S_{V_1\to V_2}$ 始终保持为零。因此,当信号入射到腔模 a_1 时,只有信号的 U 分量可以从腔模 a_2 透射且透射率大于 1($|S_{U_1\to U_2}|^2>1$),但是入射到腔模 a_2 的信号不能从腔模 a_1 透射出。因此,腔模 a_1 和 a_2 之间的放大是单向的且是相位敏感的。此外,放大的方向可以通过调节相位差 ϕ 为 $\pi/2$ 进行改变,如图 6.3.2(b) 所示。我们发现 $\omega=0$ 时可以实现从 V_2 分量到 V_1 分量的单向放大,且 $S_{U_1\to U_2}(0)=S_{V_1\to V_2}(\omega)=S_{U_2\to U_1}(\omega)=0$。

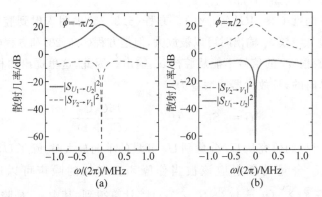

图 6.3.2　相位差 ϕ 分别等于 (a)$-\pi/2$ 和 (b)$\pi/2$ 时散射几率 $|S_{U_1\to U_2}|^2$ 和 $|S_{V_2\to V_1}|^2$ 随探测失谐量 ω 变化的曲线

(请扫 II 页二维码看彩图)

为了能更加清楚地看到探测失谐量 ω 和相位差 ϕ 对散射几率的影响,在图 6.3.3 画了 $|S_{U_1\to U_2}|^2$ 和 $|S_{V_2\to V_1}|^2$ 随 ω 和 ϕ 变化的情况。图 6.3.3(a) 表明当 $\omega=0$ 时 $|S_{U_1\to U_2}|^2$ 在 $\phi=-\pi/2$ 处出现最大值而在 $\phi=\pi/2$ 处出现最小值。但是,当 $\omega=0$ 时 $|S_{V_2\to V_1}|^2$ 在 $\phi=-\pi/2$ 处出现最小值而在 $\phi=\pi/2$ 处出现最大值。如果考虑输入信号与腔场共振的情况($\omega=0$),从图 6.3.3(b) 可以清

楚地看出散射几率$|S_{U_1 \to U_2}|^2$和$|S_{V_2 \to V_1}|^2$对相位差的依赖关系。只有当规范不变的相位和是π的整数倍时,时间-反演对称性才是完整的[31,62]。图 6.3.3(b)显示当$\phi=0$和$\pm\pi$时$|S_{U_1 \to U_2}|^2$和$|S_{V_2 \to V_1}|^2$相等。否则,如果$\phi \neq n\pi$(n是整数),光力诱导的时间-反演对称性破缺会导致非互易性传输的出现。当$-\pi<\phi<0$时$|S_{U_1 \to U_2}|^2>|S_{V_2 \to V_1}|^2$,但是当$0<\phi<\pi$时$|S_{V_2 \to V_1}|^2>|S_{U_1 \to U_2}|^2$。尤其是,当$\phi=-\pi/2$时,$|S_{U_1 \to U_2}|^2 \approx 20$ dB,$|S_{V_2 \to V_1}|^2=0$;而当$\phi=\pi/2$时,$|S_{V_2 \to V_1}|^2 \approx 20$ dB,$|S_{U_1 \to U_2}|^2=0$,因此可以实现相位敏感的单向放大。

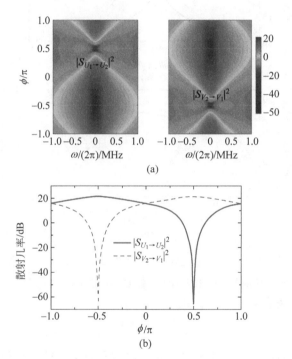

图 6.3.3 散射几率$|S_{U_1 \to U_2}|^2$和$|S_{V_2 \to V_1}|^2$随(a)相位差ϕ和探测失谐量ω以及(b)相位差ϕ(取$\omega=0$)变化的情况。其他参数与图 6.3.2 相同

(请扫Ⅱ页二维码看彩图)

此外,方程(6.3.29)表明相位敏感的放大器的增益依赖于光力协同性C_1和C_2。在图 6.3.4 中,研究了光力协同性对放大器增益和带宽的影响。图 6.3.4(a)给出了光力协同性$C_2=\{1,5,10,50\}$,$C_1=C_2^2$时散射几率

$|S_{U_1 \to U_2}|^2$ 随探测失谐量 ω 变化的情况。从图中可以看出,随着光力协同性 C_1 和 C_2 的增加,放大器的增益 $|S_{U_1 \to U_2}(0)|^2$ 和带宽都单调变大。为了能看得更加清楚,在图 6.3.4(b) 中画了增益 $|S_{U_1 \to U_2}(0)|^2$ 和带宽随光力协同性 C_2 变化的情况并保持 $C_1 = C_2^2$。相位敏感的放大器的增益和带宽随着光力协同性都单调增加,意味着这里的放大器的增益-带宽积不受限制。

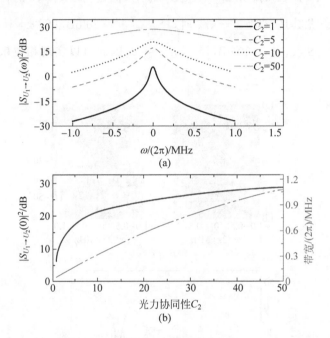

图 6.3.4 散射几率 $|S_{U_1 \to U_2}|^2$ 随(a)探测失谐量 ω 变化而光力协同性 C_2 取几个不同的值,(b)随光力协同性 C_2 变化而 $\omega=0$ 的曲线。其他参数除了 $\phi=-\pi/2$ 外与图 6.3.2 相同

(请扫Ⅱ页二维码看彩图)

增加噪声的影响是设计实用的放大器时需要考虑的一个很重要的因素。图 6.3.5 给出了放大器增加的噪声 $N_{2,\text{add}}$ 随探测失谐量 ω(其中 $C_2=\{1,5,10,50\}$)(a)和光力协同性 C_2(其中 $\omega=0$)(b)变化的情况。这里假定腔模 a_k ($k=1,2,3$)中热光子占据数为零,原因是光频区域 $\hbar\omega_k/k_B T_r \gg 1$,其中 k_B 是玻耳兹曼常数,而 T_r 是热库的温度。图 6.3.5(a)表明共振处增加的噪声可以通过增加光力协同性 $C_2(C_1)$ 而进行强烈抑制。从插图可以看出当 $C_2=$

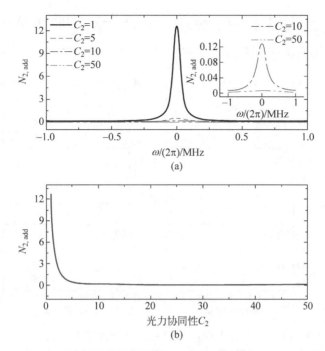

图 6.3.5 腔模 a_2 的输出端增加的噪声 $N_{2,\mathrm{add}}$ 随(a)探测失谐量 ω 变化而光力协同性 C_2 取几个不同的值(b)光力协同性 C_2 变化而 $\omega=0$ 的曲线。其他参数除了 $n_1=n_2=n_3=0, n_m=50$ 外与图 6.3.4 相同

(请扫Ⅱ页二维码看彩图)

$50, C_1=250$ 时增加的噪声接近零。此外,图 6.3.5(b)给出了增加的噪声 $N_{2,\mathrm{add}}$ 随光力协同性变化的情况,也表明随着光力协同性的增加,增加的噪声单调减小,且 C_2 增加到一定程度时增加的噪声可接近零。因此,这里提出的相位敏感的放大器可以达到增加噪声的量子极限值——零。

最后,我们在图 6.3.6 中研究腔模 $a_1(a_2)$ 和 a_3 的不同分量之间的散射情况。当相位差 $\phi=-\pi/2$ 时,图 6.3.6(a)表明入射到腔模 a_3 的信号的 U 分量可以单向放大到从腔模 a_1 输出的信号的 V 分量。此外,我们已经得到入射到腔模 a_1 的信号的 U 分量可以单向放大到从腔模 a_2 输出的信号的 U 分量,如图 6.3.2 所示。从图 6.3.6(a)和(b)还可以看出,当 $\omega=0$ 时 $|S_{U_2 \to V_3}|^2 = |S_{V_2 \to U_3}|^2 = 1$。因此,腔模 a_1、a_2、a_3 之间的相位敏感的单向放大可以通过两个增益(G)过程和一个转化(C)过程实现。当 $\phi=-\pi/2$ 时,单向放大沿着以

下路径：$U_1 \xrightarrow{G} U_2, U_2(V_2) \xrightarrow{C} V_3(U_3), U_3 \xrightarrow{G} V_1$。如果相位差 ϕ 调节为 $\pi/2$，放大的方向将改变为：$U_1 \xrightarrow{G} V_3, U_3(V_3) \xrightarrow{C} V_2(U_2), V_3 \xrightarrow{G} V_1$。

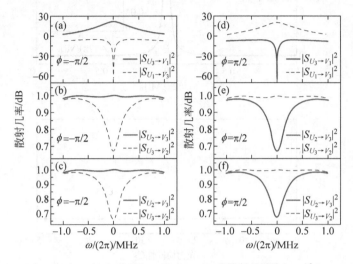

图 6.3.6 (a)~(c) $\phi=-\pi/2$ 和 (d)~(f) $\phi=\pi/2$ 时散射几率 $|S_{i \to j}|^2$ ($i,j=U_1, V_1, U_2, V_2, U_3, V_3$) 随探测失谐量 ω 变化的曲线。其他参数与图 6.3.2 相同

(请扫 II 页二维码看彩图)

6.3.4 小结

本节提出了一种基于三腔光力系统实现微波和光学光子之间实现相位敏感的单向放大器的方案[63]。研究表明，微波腔的双色驱动可以引起相位敏感的放大，而放大的方向依赖于有效光力耦合强度之间的相位差。我们还发现该光力系统中腔模之间相位敏感的单向放大现象包括两个增益过程和一个转化过程。此外，这里提出的放大器是量子极限的并且增益-带宽积没有限制，在混杂量子网络中有着潜在的应用。

参考文献

[1] CAVES C M. Quantum limits on noise in linear amplifiers[J]. Phys. Rev. D, 1982, 26: 1817.

[2] SHOJI Y, MIZUMOTO T. Magneto-optical nonreciprocal devices in silicon

photonics[J]. Sci. Technol. Adv. Mater., 2014, 15: 014602.
[3] POTTON R J. Reciprocity in optics[J]. Rep. Prog. Phys., 2004, 67: 717.
[4] APLET L J, CARSON J W. A Faraday effect optical isolator[J]. Appl. Opt., 1964, 3: 544.
[5] BI L, HU J, JIANG P, et al. On-chip optical isolation in monolithically integrated non-reciprocal optical resonators[J]. Nat. Photon., 2011, 5: 758.
[6] LIRA H, YU Z, FAN S, et al. Electrically driven non-reciprocity induced by interband photonic transition on a silicon chip[J]. Phys. Rev. Lett., 2012, 109: 033901.
[7] FANG K, YU Z, FAN S. Photonic Aharonov-Bohm effect based on dynamic modulation[J]. Phys. Rev. Lett., 2012, 108: 153901.
[8] PENG B, ÖZDEMIR Ş K, LEI F, et al. Parity-time-symmetric whispering-gallery microcavities[J]. Nat. Phys., 2014, 10: 394.
[9] CHANG L, JIANG X, HUA S, et al. Parity-time symmetry and variable optical isolation in active-passive-coupled microresonators[J]. Nat. Photon., 2014, 8: 524.
[10] HUANG R, MIRANOWICZ A, LIAO J Q, et al. Nonreciprocal photon blockade[J]. Phys. Rev. Lett., 2018, 121: 153601.
[11] FLEURY R, SOUNAS D L, SIECK C F, et al. Sound isolation and giant linear non-reciprocity in a compact acoustic circulator[J]. Science, 2014, 343: 516.
[12] ESTEP N A, SOUNAS D L, SORIC J, et al. Magnetic-free non-reciprocity and isolation based on parametrically modulated coupled-resonator loops[J]. Nat. Phys., 2014, 10: 923.
[13] WANG D W, ZHOU H T, GUO M J, et al. Optical diode made from a moving photonic crystal[J]. Phys. Rev. Lett., 2013, 110: 093901.
[14] SLIWA K M, HATRIDGE M, NARLA A, et al. Reconfigurable Josephson circulator/directional amplifier[J]. Phys. Rev. X, 2015, 5: 041020.
[15] LECOCQ F, RANZANI L, PETERSON G A, et al. Nonreciprocal microwave signal processing with a field-programmable Josephson amplifier[J]. Phys. Rev. Appl., 2017, 7: 024028.
[16] METELMANN A, CLERK A A. Non-reciprocal photon transmission and amplification via reservoir engineering[J]. Phys. Rev. X, 2015, 5: 021025.
[17] METELMANN A, CLERK A A. Nonreciprocal quantum interactions and devices via autonomous feedforward[J]. Phys. Rev. A, 2017, 95: 013837.
[18] ASPELMEYER M, KIPPENBERG T J, MARQUARDT F. Cavity optomechanics[J]. Rev. Mod. Phys., 2014, 86: 1391.
[19] MARQUARDT F, GIRVIN S M. Optomechanics[J]. Physics, 2009, 2: 40.
[20] XIONG H, SI L G, LV X Y, et al. Review of cavity optomechanics in the weak-coupling regime: from linearization to intrinsic nonlinear interactions[J]. Sci. China Phys. Mech. Astron., 2015, 58: 1.
[21] AGARWAL G S, HUANG S. Electromagnetically induced transparency in mechanical effects of light[J]. Phys. Rev. A, 2010, 81: 041803.

[22] WEIS S, RIVIÈRE R, DELÉGLISE S, et al. Optomechanically induced transparency[J]. Science, 2010, 330: 1520.

[23] SAFAVI-NAEINI A H, ALEGRE T P M, CHAN J, et al. Electromagnetically induced transparency and slow light with optomechanics [J]. Nature, 2011, 472: 69.

[24] XIONG H, SI L G, ZHENG A S, et al. Higher-order sidebands in optomechanically induced transparency[J]. Phys. Rev. A, 2012, 86: 013815.

[25] JING H, ÖZDEMIR Ş K, GENG Z, et al. Optomechanically-induced transparency in parity-time-symmetric microresonators[J]. Sci. Rep., 2015, 5: 9663.

[26] HOCKE F, ZHOU X, SCHLIESSER A, et al. Electromechanically induced absorption in a circuit nano-electromechanical system[J]. New J. Phys., 2012, 14: 123037.

[27] SINGH V, BOSMAN S J, SCHNEIDER B H, et al. Optomechanical coupling between a multilayer graphene mechanical resonator and a superconducting microwave cavity[J]. Nat. Nanotechnol., 2014, 9: 820.

[28] MASSEL F, HEIKKILÄ T T, PIRKKALAINEN J M, et al. Microwave amplification with nanomechanical resonators[J]. Nature, 2011, 480: 351.

[29] OCKELOEN-KORPPI C F, DAMSKÄGG E, PIRKKALAINEN J M, et al. Low-noise amplification and frequency conversion with a multiport microwave optomechanical device[J]. Phys. Rev. X, 2016, 6: 041024.

[30] TÓTH L D, BERNIER N R, NUNNENKAMP A, et al. A dissipative quantum reservoir for microwave light using a mechanical oscillator[J]. Nat. Phys., 2017, 13: 787.

[31] RANZANI L, AUMENTADO J. Graph-based analysis of nonreciprocity in coupled-mode systems[J]. New J. Phys., 2015, 17: 023024.

[32] HAFEZI M, RABL P. Optomechanically induced non-reciprocity in microring resonators[J]. Opt. Express, 2012, 20: 7672.

[33] SHEN Z, ZHANG Y L, CHEN Y, et al. Experimental realization of optomechanically induced non-reciprocity[J]. Nat. Photon., 2016, 10: 657.

[34] XU X W, LI Y. Optical nonreciprocity and optomechanical circulator in three-mode optomechanical systems[J]. Phys. Rev. A, 2015, 91: 053854.

[35] RUESINK F, MIRI M A, ALÙ A, et al. Nonreciprocity and magnetic-free isolation based on optomechanical interactions [J]. Nat. Commun., 2016, 7: 13662.

[36] MIRI M A, RUESINK F, VERHAGEN E, et al. Optical nonreciprocity based on optomechanical coupling[J]. Phys. Rev. Appl., 2017, 7: 064014.

[37] SOHN D B, KIM S, BAHL G. Time-reversal symmetry breaking with acoustic pumping of nanophotonic circuits[J]. Nat. Photon., 2018, 12: 91.

[38] TIAN L, LI Z. Nonreciprocal quantum-state conversion between microwave and optical photons[J]. Phys. Rev. A, 2017, 96: 013808.

[39] XU X W, LI Y, CHEN A X, et al. Nonreciprocal conversion between microwave

and optical photons in electro-optomechanical systems[J]. Phys. Rev. A, 2016, 93: 023827.

[40] XU X W, CHEN A X, LI Y, et al. Nonreciprocal single-photon frequency converter via multiple semi-infinite coupled-resonator waveguides[J]. Phys. Rev. A, 2017, 96: 053853.

[41] RUESINK F, MATHEW J P, MIRI M A, et al. Optical circulation in a multimode optomechanical resonator[J]. Nat. Commun., 2018, 9: 1798.

[42] PETERSON G A, LECOCQ F, CICAK K, et al. Demonstration of efficient nonreciprocity in a microwave optomechanical circuit[J]. Phys. Rev. X, 2017, 7: 031001.

[43] BERNIER N R, TÓTH L D, KOOTTANDAVIDA A, et al. Nonreciprocal reconfigurable microwave optomechanical circuit [J]. Nat. Commun., 2017, 8: 604.

[44] BARZANJEH S, WULF M, PERUZZO M, et al. Mechanical on-chip microwave circulator[J]. Nat. Commun., 2017, 8: 953.

[45] FANG K, LUO J, METELMANN A, et al. Generalized non-reciprocity in an optomechanical circuit via synthetic magnetism and reservoir engineering[J]. Nat. Phys., 2017, 13: 465.

[46] MALZ D, TÓTH L D, BERNIER N R, et al. Quantum-limited directional amplifiers with optomechanics[J]. Phys. Rev. Lett., 2018, 120: 023601.

[47] LI Y, HUANG Y Y, ZHANG X Z, et al. Optical directional amplification in a three-mode optomechanical system[J]. Opt. Express, 2017, 25: 18907.

[48] ZHANG X Z, TIAN L, LI Y. Optomechanical transistor with mechanical gain[J]. Phys. Rev. A, 2018, 97: 043818.

[49] JIANG C, SONG L N, LI Y. Directional amplifier in an optomechanical system with optical gain[J]. Phys. Rev. A, 2018, 97: 053812.

[50] DE LÈPINAY L M, DAMSKÄGG E, OCKELOEN-KORPPI C F, et al. Realization of directional amplification in a microwave optomechanical device[J]. Phys. Rev. Appl., 2019, 11: 034027.

[51] SHEN Z, ZHANG Y L, CHEN Y, et al. Reconfigurable optomechanical circulator and directional amplifier[J]. Nat. Commun., 2018, 9: 1797.

[52] BARZANJEH S, AQUILINA M, XUEREB A. Manipulating the flow of thermal noise in quantum devices[J]. Phys. Rev. Lett., 2018, 120: 060601.

[53] SEIF A, DEGOTTARDI W, ESFARJANI K, et al. Thermal management and non-reciprocal control of phonon flow via optomechanics[J]. Nat. Commun., 2018, 9: 1207.

[54] XU H, JIANG L, CLERK A A, et al. Nonreciprocal control and cooling of phonon modes in an optomechanical system[J]. Nature, 2019, 568: 65.

[55] WOLLMAN E E, LEI C U, WEINSTEIN A J, et al. Quantum squeezing of motion in a mechanical resonator[J]. Science, 2015, 349: 952

[56] LECOCQ F, CLARK J B, SIMMONDS R W, et al. Quantum nondemolition

measurement of a nonclassical state of a massive object[J]. Phys. Rev. X, 2015, 5: 041037.

[57] LEI C U, WEINSTEIN A J, SUH J, et al. Quantum nondemolition measurement of a quantum squeezed state beyond the 3 dB limit[J]. Phys. Rev. Lett., 2016, 117: 100801.

[58] PIRKKALAINEN J M, DAMSKÄGG E, BRANDT M, et al. Squeezing of quantum noise of motion in a micromechanical resonator[J]. Phys. Rev. Lett., 2015, 115: 243601.

[59] OCKELOEN-KORPPI C F, DAMSKÄGG E, PIRKKALAINEN J M, et al. Noiseless quantum measurement and squeezing of microwave fields utilizing mechanical vibrations[J]. Phys. Rev. Lett., 2017, 118: 103601.

[60] GARDINER C W, COLLETT M J. Input and output in damped quantum systems: quantum stochastic differential equations and the master equation[J]. Phys. Rev. A, 1985, 31: 3761.

[61] CLERK A A, DEVORET M H, GIRVIN S M, et al. Introduction to quantum noise, measurement, and amplification[J]. Rev. Mod. Phys., 2010, 82: 1155.

[62] KOCH J, HOUCK A A, HUR K L, et al. Time-reversal-symmetry breaking in circuit-QED-based photon lattices[J]. Phys. Rev. A, 2010, 82: 043811.

[63] JIANG C, SONG L N, LI Y. Directional phase-sensitive amplifier between microwave and optical photons[J]. Phys. Rev. A, 2019, 99: 023823.